La Niña and its impacts

La Niña and its impacts:
Facts and speculation

Edited by Michael H. Glantz

**United Nations
University Press**

TOKYO · NEW YORK · PARIS

United Nations University Press
The United Nations University, 53-70, Jingumae 5-chome,
Shibuya-ku, Tokyo, 150-8925, Japan
Tel: +81-3-3499-2811   Fax: +81-3-3406-7345
E-mail: sales@hq.unu.edu
http://www.unu.edu

United Nations University Office in North America
2 United Nations Plaza, Room DC2-2058-2068, New York, NY 10017, USA
Tel: +1-212-963-6387   Fax: +1-212-371-9454
E-mail: unuona@igc.apc.org

United Nations University Press is the publishing division of the United Nations University.

Cover design by Joyce C. Weston
Cover image by Climate Prediction Center

Printed in the United States of America

UNUP-1071
ISBN 92-808-1071-5

Library of Congress Cataloging-in-Publication Data

La Niña and its impacts : facts and speculation / edited by Michael H. Glantz.
    p.   cm.
Includes bibliographical references and index.
ISBN
1. La Niña Current. 2. La Niña Current—Environmental aspects.
I. Glantz, Michael H.
GC296.8.L3   L3   2002
551.6—dc21   2002002785

Contents

List of tables and figures

Tables

Figures

Figures followed by (*c*) are in color, grouped together in the center pages

Preface

In July 1998, the National Center for Atmospheric Research[1] (Boulder, Colorado) convened a workshop focused on the phenomenon known as La Niña. It was called the La Niña Summit for two reasons: admittedly, the workshop title was purposely used to draw attention to the meeting. It brought together as many El Niño experts as possible, given limited funding. It was also convened as a parallel to the October 1997 El Niño Summit, which had been held in California by the US government's Federal Emergency Management Agency (FEMA). While the real reason behind FEMA's El Niño Summit was mainly political (for a variety of reasons, such as early warning; public education; to heighten interest in the forthcoming global warming conference in Kyoto, Japan; an attempt to spark the undertaking of preventive measures, for reasons related to California state and federal politics), it did serve to highlight the fact that El Niño is a recurring and potentially devastating phenomenon, at least to Californians.

The true catalyst for convening a La Niña Summit was to identify the state of scientific understanding of La Niña, the name given to anomalously cold sea surface temperatures in the central and eastern tropical Pacific (it is also referred to as a "cold event"). Researchers and the media had increasingly made reference to forecasts of an intense La Niña event, as the 1997–98 El Niño slowly began to fade away in early 1998 until its sharp decay in May. Statements appeared about how La Niña was the opposite of El Niño and that La Niña's impacts around the globe would be the opposite to those just felt as a result of the most intense El Niño of the century in 1997–98.

Interviews with climate scientists and El Niño researchers that appeared in the electronic and printed media provided wide-ranging speculation about the impacts that might be expected to accompany an intense La Niña event. Previous La Niñas and their impacts were cited as providing possible analogous situations: 1988–89 and 1995–96. Recall that the intense drought in the US Midwest in 1988 was linked by some scientists to the moderate La Niña event that was in progress at that time. Yet, despite the flurry of interest in La Niña in mid-1998, the major emphasis of scientific research over the past few decades has primarily focused on El Niño and not on the cold counterpart of the sea surface temperature oscillation in the central and eastern equatorial Pacific (called El Niño/Southern Oscillation). It was as if, for the public, the media, and policy makers, La Niña as well as normal periods provided an interlude between the more frequent, and seemingly more devastating, El Niño.

Recently, however, it has been suggested that, for North America for example, the impacts of a La Niña event on various sectors of society are more costly than for El Niño (Changnon, 2000). Although such an assessment was preliminary, it opened up an interesting research question for those countries affected either negatively or positively by ENSO's warm and cold extremes. La Niña, as a natural phenomenon that can affect human activities in various ways, has become a concern to societies around the globe. Forecasts and assessments of La Niña and its impacts are no longer just of passive interest, as people wait to hear about the next El Niño. No longer is it necessary for societies to wait for 4.5 years on average in order to be reminded about how changes in sea surface temperatures in the tropical Pacific can affect them. La Niña forecasts also provide usable information for decision makers in many socio-economic sectors of society. Such information would enable societies to reduce their vulnerabilities to regional climate and climate-related anomalies (i.e., drought, floods, frost, fire, cyclonic activity) that are likely to accompany La Niña.

The chapters that follow have been placed into sections categorized by theme. Part 1 presents an introduction to La Niña – the cold extreme of the ENSO cycle, an overview of the La Niña process, a brief description of the El Niño and La Niña events of 1997–2000 and definitions of La Niña. Part 2 provides the reader with a glimpse of the state of scientific knowledge about cold events. Experts present their views on what constitutes "normal" in the context of the tropical Pacific, on teleconnections (or the effects of La Niña on regional climates around the globe), how climate change might affect the behavior of ENSO extremes, the degree of symmetry between El Niño and La Niña with regard to physical and societal impacts, the difficulties for attributing specific distant meteorological anomalies to a La Niña event, the forecasting of La Niña's onset, comparing the forecasting of ENSO's extremes, and a comparison of the monitoring needs for La Niña as compared to those for El Niño.

Part 3 provides case studies of the expected impacts associated with La Niña events in different countries and in different socio-economic sectors. Brief case studies were drawn from North America, Latin America, sub-Saharan Africa, and Asia. Sectors discussed include agriculture, water, energy, food, health, and fisheries.

The role of the media as educators and as conduits to the public of scientific information during recent El Niño and La Niña events is presented in Part 4. This section contains papers on the interaction between the media and scientists in general and about El Niño in particular, using examples drawn from the USA, Australia, and Japan. Also included is a discussion by a panel of media representatives, which was held at the La Niña Summit.

Part 5, entitled "More thoughts on La Niña," provides comments on researching the physical and societal aspects of the phenomenon and about forecasting climate. Although much more could have been presented about La Niña, limitations of time, space, and resources restricted us to topics in this volume. It is our hope, however, that each country and even that each weather-sensitive sector within a country will undertake an assessment of how its activities can be affected by either extreme of the ENSO cycle. It is also important for the citizens of these countries to become more familiar with La Niña and El Niño phenomena. Total coverage of the La Niña Summit was carried live (in real time) on the Internet by San Francisco's Exploratorium. Summit discussions are still available at the following website (www.esig.ucar.edu/lanina).[2]

Michael H. Glantz

Notes

1. The National Center for Atmospheric Research is sponsored by the National Science Foundation.
2. Figures in color are indicated by (*c*) following the first instance in the text, and are grouped together in the center of the book.

REFERENCE

Changnon, S.A. (ed.), 2000: *El Niño 1997–98: Climate Event of the Century*. New York: Oxford University Press.

Acknowledgments

The person most responsible for the production of this book, the first devoted solely to the phenomenon known as La Niña, is D. Jan Stewart. She worked tirelessly, helping to put together the La Niña Summit in July 1998 that brought together El Niño experts to share their insights about an air – sea interaction previously viewed as being of little importance to science or to society. She then shepherded through to publication the papers of the Summit contributors and some additional ones that were prepared after the Summit. Her efforts to produce a solid publication were once again successful.

A book of this kind is really the product of the labor of many people, not the least among them the contributors and the participants to the Summit itself. The diligent efforts of research assistants such as Autumn Moss were invaluable. The administrative support for a variety of aspects of the manuscript was done with great interest and care by Anne Oman in the Environmental and Societal Impacts Group.

Special thanks go to UNU Program Officer Zafar Adeel, who saw the value of making La Niña a household word around the globe, given that its impacts can be as devastating as those of the better-known El Niño. The United Nations University Press people, such as Janet Boileau, Gareth Johnston, and Yoko Kojima have contributed invaluable support to this publication.

And not the least of all, my warmest thanks are given to my wife, Karen Lynch, who has had to listen to a lifetime's worth of facts about La Niña (and El Niño) in just a couple of decades. What patience.

Michael H. Glantz
Boulder, Colorado
March 2002

Part I

Introduction to La Niña

La Niña: An overview of the process

Michael H. Glantz

Background

For millennia, people have tried to understand, predict, forecast, and guess what might be the natural variations of local and regional climate on seasonal and on year-to-year time scales. Among the earliest recorded treatises on weather and climate was Aristotle's *Meteorologica*, written in the fifth century BC. At least since then, seasonal and interannual variations have been a major concern to leaders and followers alike for many reasons, including food production, water resources management, shelter, and public safety. Other reasons for concern about climate and weather include general interest, curiosity, and an ever-present human desire to foresee the future. It was only toward the end of the nineteenth century that the forecasting of weather and climate started to become a growth research industry. Societal interests and concerns have been discussed by Hubert Lamb (1977) and Reid Bryson and Thomas Murray (1977), among many others.

Following a devastating famine in India in the late 1870s, interest in forecasting the behavior of the Asian monsoon received considerable attention from the British. How to predict the behavior of the Indian monsoon sparked numerous studies by various researchers who attempted to forecast the breakdown of the rain-bearing South Asian monsoon, in order to avert food shortages and famines on the Indian subcontinent, at the time a colony in the British Empire. Some researchers focused on the

3

possibility that an understanding of temperature anomalies in the tropical Pacific and Indian Oceans was the key to the improved forecasting of potential food production problems in India.

During the next few decades, researchers such as Charles Todd, Henry Blanford, J.N. Lockyer and Hugo Hildebrandson searched for reasons behind India's monsoon failure. In the earliest decades of the twentieth century, Gilbert Walker studied changes in atmospheric pressure at sea level in the western and central Pacific Ocean and in the Indian Ocean. He identified a seesaw-like process, naming it the Southern Oscillation. Scientists have since developed a Southern Oscillation Index, which has served as a statistical measure to identify the likely onset of an El Niño or a La Niña event, when sea level pressure is low in Darwin, Australia (i.e., rainy conditions prevail), it is usually high in Tahiti (i.e., dry conditions prevail), and vice versa.

The way that many of Walker's peers in the meteorological profession looked at his teleconnections work was captured in his obituary that appeared in 1959 in the *Quarterly Journal of the Royal Meteorological Society*:

Walker's hope was presumably not only to unearth relations useful for forecasting, but to discover sufficient and sufficiently important relations to provide a productive starting point for a theory of world weather. It hardly seems to be working out like that.

Meanwhile, in the early 1890s on the eastern side of the Pacific, Peruvian geographers began to write about a shift every few years or so in their coastal ocean current that brought heavy rainfall to coastal desert areas in northern Peru and affected navigation, the abundance and types of fish caught, and increased vegetative growth along the usually arid Peruvian coast. At a conference in Lima (Peru) in 1892, Peruvian geographers referred to (and perhaps named) this phenomenon as El Niño, apparently for the first time. In earlier decades in the 1800s and in preceding centuries, however, Peruvians would refer to the "septennial" torrential rains that seemed to reappear every seven years or so (Sears, 1895). Septennial rains were most likely a reference to El Niño-related periodic heavy rains and destructive flooding in northern Peru.

In the late 1960s, UCLA Professor Jakob Bjerknes, using data collected during the 1957–58 International Geophysical Year (IGY) and existing research produced by others on Peru's coastal currents, identified the physical mechanisms linking sea surface temperature changes in the eastern equatorial Pacific off the coast of Peru with sea level pressure changes in the western Pacific. This basin-wide process of air-sea interaction along the equator, as opposed to the local air-sea interactions off the coast of Peru, became referred to by two meteorologists (Rasmusson and Carpenter, 1982) in the early 1980s as ENSO, the El Niño/Southern Oscillation. It is now recognized that ENSO's warm and cold extremes affect weather and climate systems around the globe.

It was not long before Bjerknes's hypothesis was successfully tested by nature, with the onset of the strong 1972–73 El Niño event. With each event since then, the scientific community has become increasingly interested in and more knowledgeable about the phenomenon. Although the 1972–73 El Niño was devastating in its own right, especially to Peru, and although it piqued the interest of some physical scientists, it was really the 1982–83 event that prompted the scientific community and a handful of governments to focus on improving their understanding and their forecasts of this phenomenon in order to prepare for its possible impacts. No researcher had publicly forecast the 1982 event well in advance of its onset and, as a result, scientists and societies around the globe were blind-sided by the timing and magnitude of this event. Because of its intensity and the high visibility and high costs of its impacts on societies and on ecosystems, it earned the label at the time as the "El Niño of the Century."

The 1997–98 El Niño was another example of an El Niño that surprised the research community, because of its rapid onset and, later, because of its unexpected intensity. After all, there had already been an El Niño of the Century. Yet, in the short time span of just 15 years, the 1982–83 event was replaced by the 1997–98 event as the El Niño of the Century. Many potential users of weather information came to regard El Niño-related information more seriously for use in their particular decision-making activities. All of this is good news to researchers, governments, and societies in general. However, during the recent period of growing interest in and concern about warm events, other parts of the ENSO cycle – normal conditions and cold events – were given much less attention.

Introducing La Niña

Figures 1-1 (*c*) and 1-2 (*c*) depict the comparisons of selected major El Niño events and of La Niña events, using a composite index, the MEI (Multivariate ENSO Index). The index was constructed by NOAA scientist Klaus Wolter, using a set of six variables, as noted in the caption for Figure 1-1.

Until the late 1990s, there had been relatively little interest worldwide in the societal impacts of La Niña. It is likely that the intense 1997–98 El Niño and its rapid demise in May 1998 sparked considerable interest in forecasts of an intense cold event to begin in the summer of 1998. Scientists forecast a strong La Niña by the end of 1998. We now know with hindsight that only a moderate, but prolonged, La Niña event did develop then. It continued into the year 2000, making it one of the longest La Niña events in recent decades.

Even so, La Niña events still fail to command the level of attention achieved by El Niño among the public, the media, or decision makers. While media coverage of El Niño in the past seldom referred to La Niña, stories about La Niña usually made reference to El Niño. By inference, then, La Niña appears to be viewed as an outgrowth of El Niño, and not in any way as its equal. At the risk of reinforcing this misconception, the best place to begin a discussion of La Niña is with reference to the 1997–98 El Niño. In fact, they are of equal importance to many societies, which tend to suffer different but equally devastating impacts associated with each extreme.

Interest in retrospective assessments of the societal impacts of various El Niño events became widespread in the final years of the 1990s, as a result of the 1997–98 El Niño. Such assessments were separately undertaken by United Nations agencies, regional organizations, humanitarian aid agencies, national governments, non-governmental organizations, industries, educational institutions, official and unofficial internet websites, as well as by decision makers in a wide range of economic and social activities around the globe. A notable global effort was the response of UN agencies to the November 1997 UN General Assembly Resolution that called for an El Niño impacts assessment. This UN review of the physical (e.g., technical) aspects of the event was the responsibility of the World Meteorological Organization (WMO), while the societal aspects were coordinated by the IDNDR (International Decade for Natural Hazard Reduction, now the International Strategy for Disaster Reduction, ISDR). Each of these UN agencies established a UN task force to carry out its contribution to the overall assessment. Although there had been minimal interaction between these two assessment processes, all groups were brought together to deliver their findings at an international conference on "A Retrospective Assessment of the 1997–98 El Niño" convened in Guayaquil, Ecuador from 9–13 November 1998 (WMO, 1999).

Another assessment of the impacts of and response strategies for the 1997–98 El Niño was undertaken as the result of a grant to the United Nations Environment Programme (UNEP) and the National Center for Atmospheric Research (NCAR) from a recently established US non-profit research organization (the UN Foundation). Case studies were undertaken in 16 countries to identify the strengths and weaknesses of societal responses to both the forecast as well as the impacts of the 1997–98 El Niño (Glantz, 2001). Countries in the study included China, Vietnam, the Philippines, Indonesia, Papua New Guinea, Bangladesh, Fiji, Mozambique, Ethiopia, Kenya, Ecuador, Peru, Costa Rica, Cuba, Paraguay, and the operations of the Panama Canal. The study teams were primarily non-governmental. That was done to objectively identify lessons for use by their respective governments. Research for each case study did, however, involve the participation of relevant government agencies. This UNEP/NCAR project, which also involved the UNU, the WMO, and the IDNDR, was designed to improve societal coping mechanisms in the face of future ENSO's extreme warm *or* cold events.

Independent research findings have shown over the past two and a half decades that climate-related extremes, such as droughts, frosts, floods, blizzards, and fires, easily capture the attention of the public, policy makers, international and non-governmental organizations and especially the media, in particular when these events are in progress. For a variety of reasons, however, sympathetic interest quickly dissipates. Such usual (and now, expectable) dissipation of media, policy maker and societal interest in post-disaster recovery brings to mind the adage, "out of sight, out of mind." A similar response occurs with regard to El Niño and La Niña events and, compared to El Niño, there appears to be an even faster drop in interest in La Niña, except perhaps among researchers.

Until the mid-1990s, many studies and the graphics within them would refer to the ENSO cycle only in terms of El Niño and non-El Niño conditions in the tropical Pacific Ocean, as shown in Figures 1-3 and 1-4. For the most part, researchers apparently did not see the need to distinguish between normal (i.e., average) and La Niña conditions. Yet, each of these conditions in the central Pacific surely has different implications for attempts to forecast climate and climate-related anomalies around the globe.

In many parts of the world, people consider El Niño as *the* climate-related "hazard" that *they* need to be concerned about (Glantz, 2001). When El Niño goes away, the cause of many of their worst climate-related problems is believed to have disappeared – at least for a while – until the next El Niño develops (or so they believe). To them it is of little, if any, concern if the sea surface temperatures (SSTs) return to a normal range in the tropical Pacific or if those SSTs move into La Niña conditions. They tend to see both of these conditions as "normal." In this regard, then, to them normal could be defined as the absence of an El Niño episode. This view has unwittingly been reinforced by the media, which tend to lose interest in El Niño as it shows signs of weakening. Then, topical political or economic issues tend to push El Niño off the front page or the TV screen.

For many other regions, in terms of anomalies spawned by the *extreme* cold phase of the ENSO cycle, what might take place is quite uncertain.

This is not necessarily because La Niña events are less predictable (as some scientists have suggested), but because fewer La Niña events have occurred since the mid-1970s relative to the previous (1950–77) period. The frequency and intensity of recent ENSO extremes are suggested in the following chart (Figure 1-3).

La Niña events were not observed in previous decades at the high level of scrutiny of the 1990s. Cold events were apparently viewed by many as providing an interlude between two potentially harmful El Niño events. Because of these factors, the societal and environmental impacts of specific La Niña events have been reconstructed using historical, anecdotal, and proxy information. Although very useful, such information becomes increasingly uncertain and unreliable as one looks further back in time.

Much of what one hears about La Niña's climate-related impacts in various locations around the globe is either speculation by media reporters or are "guesstimates" provided by knowledgeable researchers. The public is left to itself to decide which teleconnections (i.e., distant impacts) are reliable.

Today, the mere mention of the possible onset of a La Niña event will likely spark reactions from decision makers in countries where a strong (usually adverse) La Niña disruption of their regional climate is believed to exist (e.g., Philippines, Malaysia, Brunei, the United States). It is important, however, to keep the following points in mind: (1) forecasting related to La Niña or El Niño is not a perfect science, as witnessed by the forecast community's inability to correctly forecast the duration of the 1998–2000 La Niña or the onset of the 1997–98 event; (2) a drop in sea surface temperatures in the tropical Pacific from relatively high levels does not ensure that a La Niña event will immediately follow; (3) a strong La Niña event does not necessarily follow a strong El Niño; (4) cold events, like the warm ones, have varying levels of intensity, each of which is accompanied by its own combination of worldwide teleconnections.

Over the years, most scientists have focused on forecasting the onset of ENSO's warm extreme and to a much lesser extent on the onset of its cold extreme. As a result, they have had less skill at forecasting either the intensity of these events or the locations and severity of their environmental and societal impacts worldwide. This is an important point in light of efforts by various scientific groups to put the reliability of their forecast in the best light possible. For example, Barnston et al. (1999a) assessed the correctness of 15 experimental forecasts of the 1997–98 El Niño. They concluded the following:

It becomes clear on review of the various forecasts that were issued quarterly in the *ELLF Bulletin* that most of the forecasts did not identify a tendency toward

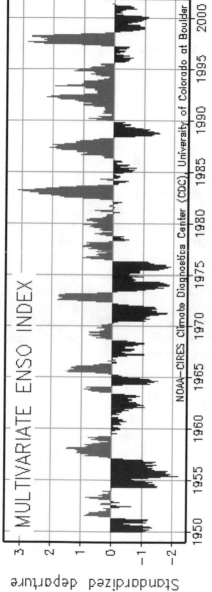

Fig. 1-3 Time series of the El Niño/Southern Oscillation using the MEI (definition of MEI: see Figure 1-1) (from www.cd.noaa.gov/~kew/MEI/).

an El Niño onset until the first quarter (i.e., March) of 1997, when rapid warming was just beginning to occur. Similarly, they did not acknowledge the appearance of an exceptionally strong event until the June 1997 forecast, when the event was becoming very strong. (p. 235)

Yet, the findings were contrary to public statements by those forecasters and modelers who reported their "forecasting successes" (Barnston et al., 1999b).

What La Niña is

Under non-El Niño conditions, a pool of warm water is usually located in the western equatorial Pacific. It provides moisture to the atmosphere through evaporative processes which lead to the formation of convective activity and rain-producing cloud systems in that region. As a result, heavy rains, considered by many in the region to be normal, provide the water resources needed in East, South and Southeast Asian countries and in the Pacific islands for agriculture, hydropower, water supplies and navigation. Meanwhile, extremely cold upwelled water along the equator in the eastern part of the Pacific and along the Peruvian coast creates atmospheric subsidence (i.e., descending motion of the atmosphere). This inhibits cloud formation and enhances the arid conditions along the western coast of South America. This represents the non-El Niño conditions referred to earlier, which to many observers include both La Niña *and* normal conditions.

As noted earlier, many of the pre-1997 diagrams that depicted the ENSO process presented only two states of air-sea interaction in the tropical Pacific – El Niño and normal (shown in Figure 1-4). In fact, many diagrams continue to show only these two stages of air-sea interaction in the region. Today, however, new diagrams are appearing that depict La Niña conditions as different from normal (Figure 1-5).

Figure 1-6 provides another way to show that there are the three different states of SST conditions in the equatorial Pacific. As one can see from Figure 1-6, La Niña and normal conditions appear to be relatively close to each other, especially to those living in the western part of the Pacific basin.

It was only at the very end of the twentieth century that the media took an interest in the La Niña phenomenon. Interestingly, La Niña headlines appear to have been much less threatening in tone than those used for El Niño stories, as suggested in Figure 1-7.

In a strict sense, La Niña exists only when relatively extreme cold sea surface temperatures appear for a designated period of time (several

Normal conditions

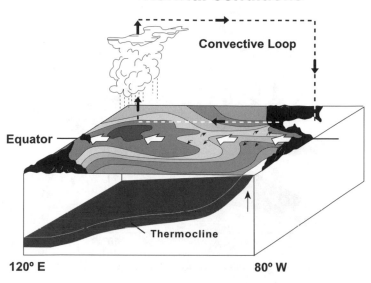

El Niño conditions

Fig. 1-4 Normal and El Niño conditions (from NOAA/PMEL/TAO Project Office, www.pmel.noaa.gov/tao/proj_over/diagrams/).

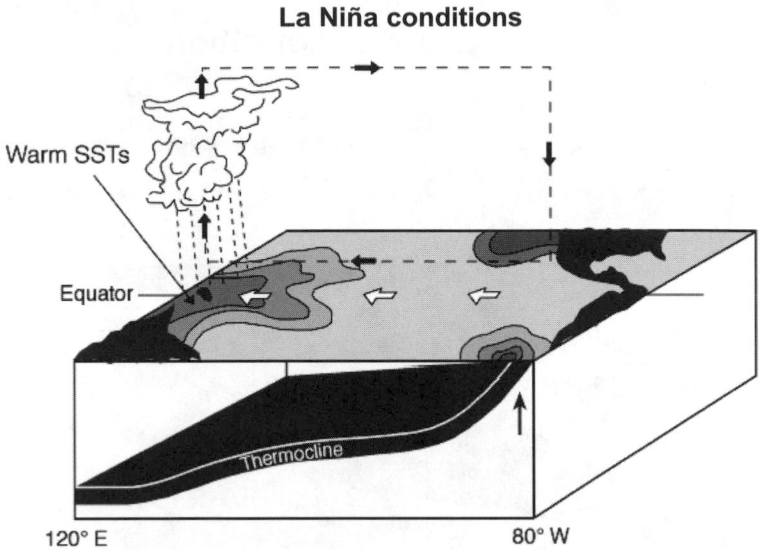

La Niña conditions

Fig. 1-5 La Niña conditions (from ESIG/NCAR).

months) in the central and eastern equatorial Pacific Ocean. This condition occurs when westward-blowing winds are strong, when the value of
the Southern Oscillation Index is highly positive, when the thermocline
in the western Pacific is depressed and in the eastern Pacific is near the
ocean's surface (as shown in Figure 1-5). La Niña events can be weak,
moderate, strong, very strong and extraordinary. Researchers have said
that the 1988–89 La Niña was strong, and that the 1995 event was a weak
one. Others have suggested that La Niña events such as the one in 1984–
85 were weak. A first comparison of the intensities of La Niña events
noted that "The sea surface temperature indices indicate that the 1998–
2000 cold episode is comparable in intensity to the cold episodes of 1970–
71 and 1973–76, is stronger than the cold episode in the mid-1980s, and
is weaker than the cold episode of 1988–89" (NCEP, 2000). Such desig-
nations are important because the environmental and societal impacts of
a cold event will depend on the level of intensity to which it develops,
other things being equal. However, there have been only a few La Niña
events observed in recent decades, making it difficult to use these categories at present with any degree of accuracy.

What La Niña does

Many of the existing climate impacts maps depicting the possible impacts
of La Niña around the globe are variations of maps produced in the

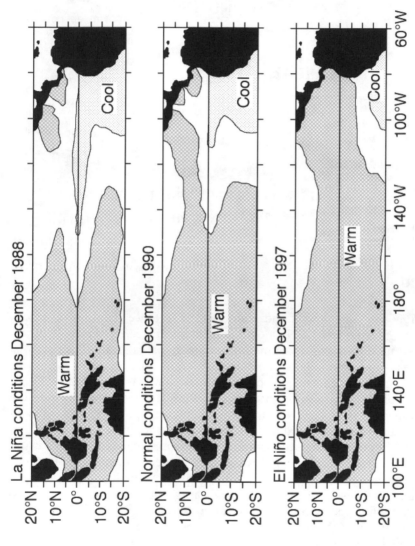

Fig. 1-6 La Niña, El Niño, and normal sea surface temperature (SST) conditions. Redrawn from NOAA's PMEL website (www.pmel.noaa.gov/tao/proj_over/diagrams/).

13

Sick of El Niño? Try La Niña

La Niña could spawn rough hurricane season

'LA NINA' IS BAKING ARIZONA
MONSOONS' FATE PONDERED

After Mild El Niño,
Brace for La Niña

HEADLINE: EL NINO'S COUNTERPART

What impact could
La Niña have on
the rest of 1998?

La Niña, Sunspot
Cycle Could Mean Drought

La Niña Blues

Some Fear Worst From Hurricane Season
Forecasters Debate Effects of La Niña's Cooling

Sibling rivalry is brewing; Exit El Niño, enter La Niña

Tired of El Niño?
Here Comes La Niña

What About La Niña?

EL NIÑO'S PESKY SISTER LURKS

El Niño's Wicked Cousin May Visit, Bringing Cold and Wet Weather

An El Niño Flip-Flop
La Niña is Warmer, Drier

Fig. 1-7 La Niña headlines taken from the international media (from ESIG/ NCAR).

mid-1980s by Ropelewski and Halpert (1987). Since then, however, additional El Niño and La Niña events have occurred. Therefore, it is necessary that the scientific community update and recalibrate those original impacts maps, especially because they are frequently used by forecasters, researchers, the media and the public for decision-making purposes. Other aspects of La Niña also require careful consideration. They include the following: the symmetry of its impacts when compared to El Niño; the attribution of climate-related impacts occurring at the time of a cold event; the strength of teleconnections; monitoring approaches to ENSO's warm and cold extremes; and global warming's influence on the ENSO cycle. This was the reason for convening the La Niña Summit in July 1998. Although briefly noted in the next few paragraphs, these issues are discussed by the experts in the chapters that follow.

An example of the attribution problem would be the following: A major drought took place in the US Midwest in the summer of 1988. This severe and most costly drought in US history (at a cost of an estimated US$40 billion) was cited by some researchers to reinforce the argument that global warming would mean an increase in similarly intense droughts in the future (e.g., Hansen, 1988). However, an equally plausible scenario, suggested by Trenberth (1988), was that the 1988 Midwest drought was attributed to La Niña conditions in the equatorial Pacific. If so, how was La Niña involved? Could one have then predicted that there was a higher probability of a similarly intense drought occurring in the same region during the 1998–2000 La Niña?

As it happened, there was a major drought in the US in the summer of 1998. However, it occurred in the US Northeast, and not in the Midwest. There is some evidence, but not nearly enough, to say with certainty that a La Niña summer in North America is likely to be hotter and drier than normal. Attribution problems are linked to scientific uncertainty and to the absence of long-term reliable scientific data. Which distant climate-related impacts might one reliably "blame" on La Niña?

Similar questions of attribution can be raised for the January 1998 devastating ice storm in the Northeast of North America. To what extent was the 1997–98 El Niño involved in its occurrence or in its severity? A recent study by Barsugli et al. (1999) did suggest that El Niño had an influence on the occurrence of this unusual and rare ice storm.

A basic concern is the importance of knowing which years or, better yet, which parts of years were truly La Niña conditions. Such basic information is needed by those hoping to develop effective strategies to prevent, mitigate, or adapt to the impacts of either of the extremes of the ENSO cycle. Physical scientists, however, appear to be less concerned about what societal impacts researchers might consider to be the imprecise labeling of specific years as either ENSO cold or warm event years. For their work, scientists rely on time series of sea surface temperatures (in one part or another of the equatorial Pacific), sea level pressure, thermocline depth and outgoing long-wave radiation. For them, these time series identify the occurrence of ENSO's extremes. While labeling specific calendar years as La Niña or El Niño may be shorthand for knowledgeable scientists, it can be misleading to non-specialists who rely on these researchers to identify and date with accuracy the onset and decay of El Niño and La Niña episodes.

Putting a cost on ENSO's impacts

Costing out both the positive and negative effects of an ENSO warm or cold extreme is admittedly not an easy task. To do it correctly involves identifying first-, second- and maybe even third-order impacts, e.g., those adverse effects that ripple through an economic or social system. Because data collection is not done using the same guidelines or according to the same standard in all countries, information about La Niña and El Niño and their impacts is often anecdotal and subjective. Hence, the estimates of the global economic impacts of El Niño are wide-ranging and, for the case of the 1997–98 El Niño, varied by a factor of three: $30 billion to $100 billion (Sponberg, 1999). As poor as data collection might be during "normal" climate disruptions, it becomes more difficult during years of natural disaster-related emergencies.

One American author recently suggested that La Niña events are more costly for the US than El Niño events (Changnon, 1999), basing this view on an evaluation of the ENSO extremes in the 1997–99 period. His assessment received widespread coverage in the popular press in the United States. Aside from reviewing the appropriateness of the methods used to gather the information used to support such a statement, it is important to note that there have been twice as many warm events as cold events since the mid-1970s. Thus, it is not yet clear whether, on an event-by-event or on a collective basis, cold events would still be considered more damaging to the US than warm ones. Furthermore, from a US national perspective, El Niño events appear to be more damaging to its West Coast states than La Niña events. La Niña events appear, however, to be more damaging to its East Coast states. So, there is a regional consideration that need to be accounted for when determining the socictal costs and benefits to a nation of either of ENSO's extremes. Such an objective assessment has not yet been done.

Physical indicators of La Niña

The following items are *some* of the characteristics or telltale signs that a La Niña event may be emerging:
- unusually cold temperature in the central and eastern equatorial Pacific Ocean
- unusually low pressures west of the date line, and high pressure east of the date line in the low-latitude Pacific
- the Southern Oscillation Index (SOI) is positive
- stronger than normal easterlies (i.e., westward-blowing trade winds)
- deep and cold water welling up to the ocean surface along the Peruvian coast and the equator in the central and eastern equatorial Pacific
- heavy rainfall over the warmer-than-normal water in the western Pacific
- a rise in sea level in the western Pacific and decline in the eastern Pacific
- a depressed thermocline in the western equatorial Pacific
- compression of convection into smaller and smaller areas in the western Pacific
- strong surface winds pushing greater amounts of warm surface water toward the western Pacific

Suggested global impacts of La Niña

A rule of thumb that has emerged is that La Niña's global impacts are generally opposite to those that accompany an El Niño. For some regions, that statement is valid. For example, droughts tend to accompany El Niño in northeastern Australia, Indonesia, and the Southern Philippines, whereas heavy rains and flooding have a higher probability of accompanying La Niña in these same locations. Another example would be the coastal zone of northern Peru which is flood-prone during El Niño but usually returns to arid conditions during normal and La Niña periods. Southern Africa tends to be drought-plagued during El Niño episodes, but flood-prone during La Niña.

As noted earlier, researchers have compiled maps depicting the type and/or the timing of impacts associated with La Niña. Clearly, the composite maps published by Ropelewski and Halpert in 1987 provided a useful "statistically based" generalization of the potential effects of ENSO warm and cold extremes. The maps that follow (Figures 1-8 and 1-9) were modified for the US Department of Agriculture in order to provide needed information about the months during which La Niña-related temperature and precipitation anomalies can be expected to occur.

Many variations of such composite maps for La Niña and El Niño anomalies have been based on the Ropelewski-Halpert maps and, therefore, should not be viewed as independently derived depictions of ENSO's potential impacts. Some impacts maps are composites of the impacts that occurred during several events, while others represent impacts during an individual calendar year designated as a year in which an ENSO extreme event occurred. These two highly visual and highly popular methods are used by researchers, governments, and the media to identify what is likely to happen during a La Niña event: (1) composite maps (Figures 1-8, 1-9 and 1-10), and (2) single-year impacts maps (Figures 1-11 and 1-12).

1. Composite maps draw generalized conclusions, based on a review of impacts for a set of La Niña events. These maps average the impacts of weak as well as strong events. Figure 1-10 represents one type of composite map for cold events (Northern Hemisphere winter and spring), as produced by the Center for Ocean-Atmospheric Prediction Studies (COAPS) at Florida State University.
2. The single-year impacts map is based on the selection of a known specific past La Niña year and the assumption that many of the climate-related impacts in that particular year could be associated with that specific event and are typical of cold events. The following figures are impacts maps for La Niña years 1974 and 1988.

The texts that accompany such graphics do not note that with composites,

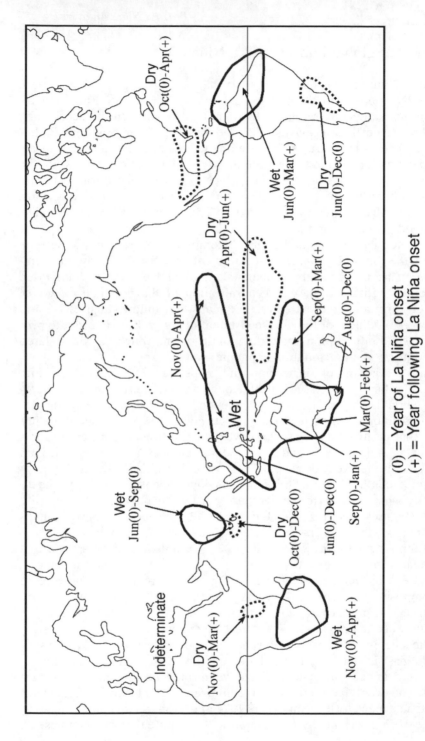

Fig. 1-8 Idealized potential rainfall impacts during La Niña events (cold episodes) (after Ropelewski and Halpert, 1987).

(0) = Year of La Niña onset
(+) = Year following La Niña onset

Dry
Oct(0)-Apr(+)

Wet
Jun(0)-Mar(+)

Dry
Jun(0)-Dec(0)

Dry
Apr(0)-Jun(+)

Wet
Nov(0)-Apr(+)

Wet

Sep(0)-Mar(+)

Aug(0)-Dec(0)

Mar(0)-Feb(+)

Sep(0)-Jan(+)

Jun(0)-Dec(0)

Dry
Oct(0)-Dec(0)

Wet
Jun(0)-Sep(0)

Indeterminate

Dry
Nov(0)-Mar(+)

Wet
Nov(0)-Apr(+)

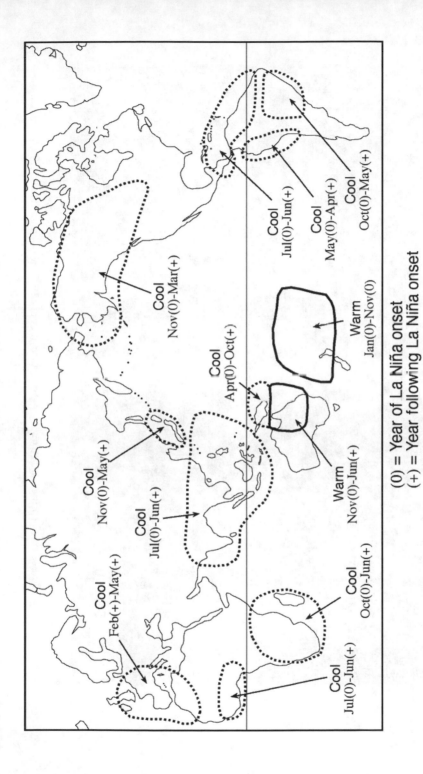

Fig. 1-9 Idealized potential temperature impacts during La Niña events (cold episodes) (after Ropelewski and Halpert, 1987).

19

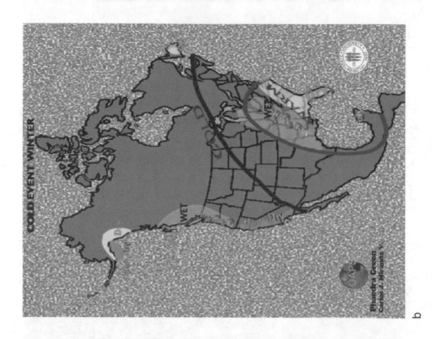

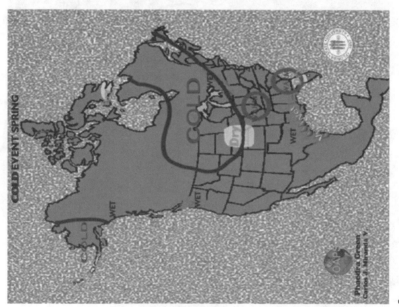

Fig. 1-10 Cold event (a) spring, and (b) winter (from Green et al., 1997, www.coaps.fsu.edu/lib/booklet/).

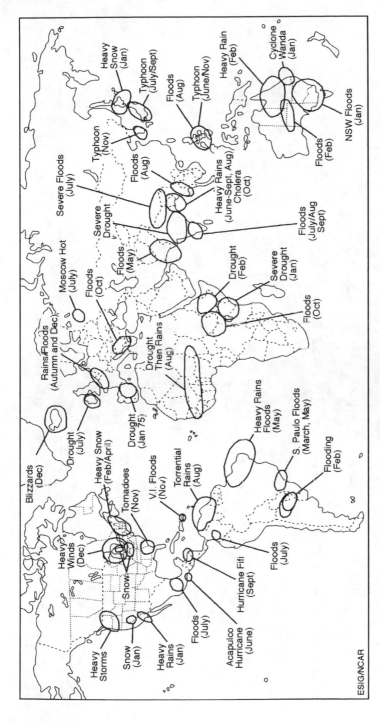

Fig. 1-11 1974 anomalies during a La Niña year. According to Trenberth (1997), a La Niña event began in June 1973 and ended in 1974, and another La Niña event began in September 1974 and ended in April 1976.

21

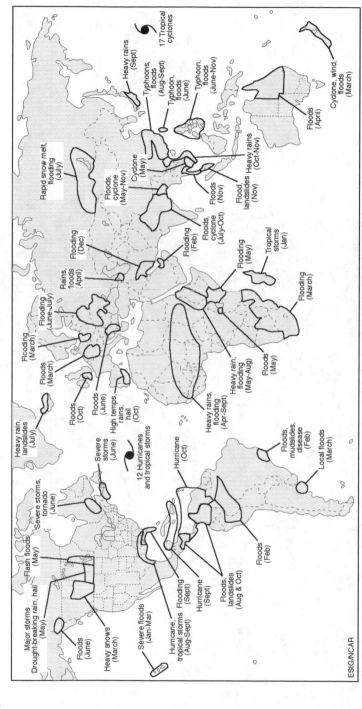

Fig. 1-12 1988 anomalies during a La Niña year. According to Trenberth (1977), this La Niña event began in May 1988 and ended in June 1989.

large events have clearer "fingerprints," i.e., attributable impacts, than the weak ones. When intense and weak events are averaged together, there is a loss of detailed information about the changes in impacts from event to event. With respect to a single-year map for a given La Niña event, the use of a single year to serve as *the* analogue for impacts resulting from cold events, while suggestive, is fraught with misinformation. While certain impacts may accompany a particular La Niña event with some degree of reliability, there is no assurance that those impacts will accompany each and every event. Furthermore, La Niña's impacts in a specific locale can either be worsened or mitigated by local and regional factors. And, because La Niña events usually straddle parts of calendar years (not so for 1999), the use of only one calendar year to identify its likely impacts would be incomplete and misleading.

REFERENCES

Barnston, A.G., M.H. Glantz, and Y. He, 1999a: Predictive skill of statistical and dynamical climate models in SST forecasts during the 1997–98 El Nino episode and the 1998 La Niña onset. *Bulletin of the American Meteorological Society*, 80(2), 217–43.

Barnston, A.G. et al., 1996b: NCEP forecasts of the El Niño of 1997–98 and its US impacts. *Bulletin of the American Meteorological Society* 80(9), 1829–1852.

Barnston, A.G., H.M. van den Dool, S.E. Zebiak, T.P. Barnett, Ming Ji, D.R. Rodenhuis, M.A. Cane, A. Leetmaa, N.E. Graham, C.R. Ropelewski, V.E. Kousky, E.A. O'Lenic, and R.E. Livezey, 1994: Long-lead seasonal forecasts – where do we stand? *Bulletin of the American Meteorological Society*, 75, 2097–114.

Barsugli, J.J., J.S. Whitaker, A.F. Loughe, P.D. Sardeshmukh, and Z. Toth, 1999: The effect of the 1997–98 El Niño on individual large-scale weather events. *Bulletin of the American Meteorological Society*, 80(7), 1399–411.

Bryson, R.A. and T.J. Murray, 1977: *Climates of Hunger: Mankind and the World's Changing Weather*. Madison, Wisconsin: University of Wisconsin Press.

Changnon, S.A., 1999: Impacts of 1997–98 El Niño-generated weather in the United States. *Bulletin of the American Meteorology Society*, 80(9), 1819–27.

Glantz, M.H. (ed.), 2001: *Once Burned, Twice Shy? Lessons Learned from the 1997–98 El Niño*. Tokyo, Japan: United Nations University Press.

Green, P.M., D.M. Legler, C.J. Miranda and J.J. O'Brien, 1997: *The North American Climate Patterns Associated with the El Niño/Southern Oscillation*. COAPS Project Report Series 97-1. Tallahassee, FL: Center for Ocean Atmosphere Prediction Studies, Florida State University; www.coaps.fsu.edu/lib/booklet/

Hansen, J.E., 1988: Quoted by Eugene Linden, Big chill for the greenhouse: Remember El Niño? Now comes its cool sibling, La Niña. *Time Magazine*, 31 October, p. 90.

Lamb, H.H., 1977: *Climatic History and the Future*. Volume 2 of *Climate: Present, Past and Future*. London: Methuen & Co. Ltd.

NCEP (National Centers for Environmental Prediction), 2000: *Climate Diagnostics Bulletin, March 2000*. Camp Springs, MD: Climate Prediction Center, National Weather Service.

Quarterly Journal of the Royal Meteorological Society, 1959: Quote from obituary of Sir Gilbert Walker, 85, 186.

Rasmusson, E.M. and T.H. Carpenter, 1982: Variations in tropical sea surface temperature and surface wind fields associated with the Southern Oscillation/El Niño. *Monthly Weather Review*, 110, 354–84.

Ropelewski, C.F. 1992: Predicting El Niño events. *Nature*, 356, 476–7.

Ropelewski, C.F. and M.S. Halpert, 1987: Global and regional scale precipitation patterns associated with the El Niño/Southern Oscillation. *Monthly Weather Review*, 115, 1606–626.

Sears, A.F., 1895: The coastal desert of Peru. *Bulletin of the American Geographical Society*, 28, 256–71.

Sponberg, K., 1999: Weathering a storm of global statistics. *Nature*, 400, 13.

Trenberth, K., 1988: Quoted in article written by Eugene Linden, Big chill for the greenhouse: Remember El Niño? Now comes its cool sibling, La Niña. *Time Magazine*, 31 October, p. 90.

Trenberth, K., 1997: The definition of El Niño. *Bulletin of the American Meteorological Society*, 78(12), 2771–7.

WMO (World Meteorological Organization), 1999: *The 1997–98 El Niño Event: A Scientific and Technical Retrospective*. WMO-No. 905. Geneva, Switzerland: WMO.

El Niño, La Niña, and the climate swings of 1997–98: A review

Michael J. McPhaden

The 1997–98 El Niño has been called "the climate event of the century." It was one of the strongest El Niños on record, with spectacular impacts on global weather variability and Pacific marine ecosystems. This El Niño abruptly terminated in mid-1998 with the rapid onset of a strong La Niña that has persisted through the end of 2000.

El Niño and La Niña comprise the warm and cold phases, respectively, of the El Niño/Southern Oscillation (ENSO) cycle, a coupled ocean-atmosphere phenomenon originating in the tropical Pacific. Here we briefly describe the evolution of the ENSO cycle during 1997–98, from the onset of the El Niño through the initial development of the La Niña. A more detailed account of events during 1997–98 can be found in Mc-Phaden (1999) and WMO (1999). Information on the subsequent evolution of the La Niña during 1999 and 2000 is available on the World Wide Web from the Climate Prediction Center (www.cpc.ncep.noaa.gov/) and the Tropical Atmosphere Ocean (TAO) array project (www.pmel.noaa.gov/tao/).

Data from the El Niño/Southern Oscillation (ENSO) Observing System (Figure 1-13 (c)) form the basis of the following presentation. Development of the ENSO Observing System, which consists of both satellite and ocean-based measurement platforms, was stimulated by the failure to forecast or even detect (until nearly at its peak) the major 1982–83 El Niño. The cornerstone of this observing system is the TAO array of moored buoys, providing oceanographic and surface meteorological data in real time (within hours of collection) for studies of year-to-year climate variations. The 1997–98 El Niño was the first El Niño for which the ENSO Observing System was in place from start to finish, having been completed in 1994 at the end of the 10-year international Tropical Ocean Global Atmosphere (TOGA) program (McPhaden et al., 1998).

The onset of the 1997–98 El Niño was heralded in the western Pacific by the appearance of episodic westerly surface winds in late 1996 and early 1997 (Figure 1-14a (c)). Shortly afterwards, sea surface temperature (SST) began to warm rapidly in the tropical eastern and central Pacific (Figure 1-14b (c)). By June-1997, SSTs reached historic highs in these regions, with SST anomalies (that is, deviations from normal) exceeding 5 °C by the end of the year. These anomalies were higher than observed at any time during the 1982–83 El Niño, which up to that time had been the strongest of the century.

As the warm SST anomalies began to wane in early 1998, prediction models were largely consistent among themselves in forecasting an end to the El Niño and the development of a La Niña sometime during the second half of 1998. The surprise to El Niño researchers, though, was how quickly the tropical Pacific switched from warm to cold conditions. The system of moored buoys detected an unprecedented 8 °C drop in sea surface temperature in the eastern Pacific from early May to early June 1998 (Figure 1-15). Such a temperature drop over a relatively brief 30-day span had never been observed before, making the termination of the El Niño as stunning as its onset.

The seeds for El Niño's demise, and for the birth of La Niña, were to be found below the sea's surface. In late 1997, the buoy array detected a cold mass of water, situated at 100–150 m depth west of the date line, which was beginning to expand eastward along the equator (Figure 1-16a (c)). This cold water progressively under-rode the layer of El Niño-warmed surface waters throughout the first part of 1998. However, because the trade winds in the eastern and central Pacific remained weak during early 1998, this cold subsurface water could not be upwelled (brought to the surface) and mixed with the warmer surface water. By late April 1998, a thin veneer of unusually warm water at the surface capped a thick layer of unusually cold water just below it across a wide swath of longitudes (Figure 1-16b (c)).

When in early May 1998 the east-to-west trade winds finally picked up to near normal strength in the eastern and central Pacific (Figure 1-14a), the cold water hovering just below the surface was drawn up to cool sea surface temperatures at a record pace. Subsequently, the patch of cold surface water expanded both eastward and westward as the La Niña developed (Figure 1-14b).

The slow development of subsurface cold-water anomalies along the

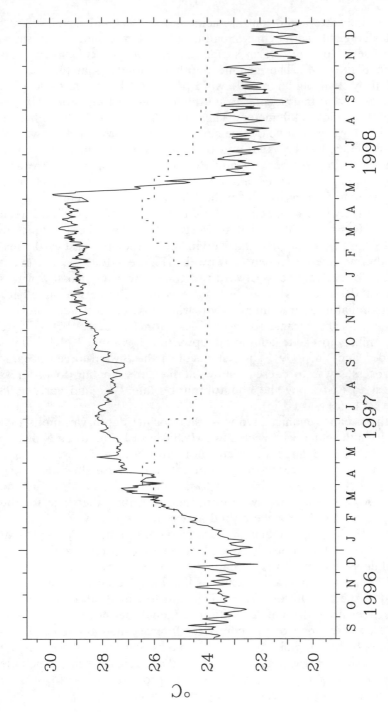

Fig. 1-15 Sea surface temperature time series on the equator at 125°W longitude. The normal monthly mean seasonal cycle at this location is shown by the dashed line.

27

equator in the eastern and central Pacific in late 1997 and early 1998 was a manifestation of thermocline shoaling. (The thermocline is the sharp boundary separating the warm surface layer from the cold interior, as shown in Figure 1-4.) Thermocline depth variability is in turn largely controlled by changing large-scale wind patterns. Changes in trade wind intensity cause the thermocline to undulate along the equator over distances of thousands of kilometers. These undulations reflect the influence of equatorial waves well below the sea's surface, which come in two important varieties, Kelvin waves and Rossby waves. Kelvin waves propagate eastward along the equator and Rossby waves propagate westward. Once generated, these waves take only a few months to cross the basin, affecting regions far removed from the winds that forced them.

As the trade winds weakened at the onset of the El Niño, wind-forced equatorial Rossby waves radiated westward causing the thermocline to shoal (upwell), and equatorial Kelvin waves radiated eastward causing the thermocline to deepen (downwell) (Figure 1-16a). As the event continued to evolve, the westward-radiating upwelling Rossby waves bounced off the land masses bordering the western Pacific and came back along the equator as upwelling Kelvin waves. Also, the trade winds in the western Pacific became stronger than normal in early 1998 (Figure 1-14a), tending to excite additional upwelling favorable Kelvin waves (McPhaden and Yu, 1999). The net result of these wind-forced oceanic wave processes was to cause a slow thermocline shoaling to progress from west to east along the equator during late 1997 and early 1998 (Figures 1-14c (c) and 1-16b).

It is the ocean dynamics involving thermocline depth variations that precondition the equatorial Pacific to switch from El Niño to La Niña and back again. These dynamics operate on relatively long time scales compared to those that govern atmospheric motions. These slow ocean dynamical processes were at work prior to the onset of the 1997–98 El Niño (McPhaden, 1999) and they were important in setting the stage for the onset of La Niña in 1998 as described above.

Many ENSO forecast models that use equations of motion to represent the dynamics of the atmosphere and ocean successfully predicted the eventual development of a La Niña in 1998. The success of these models was due in part to initialization schemes that included ocean temperature data and/or the time history of surface wind forcing, both of which can effectively constrain the slow evolution of model thermocline depth variations. However, none of the computer forecast models predicted exactly when, or how rapidly, the switch would occur from El Niño to La Niña. The transition was triggered by a rapid and relatively unpredictable strengthening of the trade winds that was the proximate cause for surface cooling in the eastern and central equatorial Pacific.

In summary, a spectacular display of climatic variability was captured in 1997–98 by a newly completed observing system designed specifically for detection and forecasting of ENSO time-scale fluctuations. This new observing system provided information at a level of detail never before possible for tracking the evolution of the ocean-atmosphere system in the tropical Pacific. Many ENSO forecast models, initialized with data from this new observing system, were correct in predicting that 1997 would be warm and the latter half of 1998 would be cold in the tropical Pacific. Moreover, large-scale warm and cold sea surface temperature anomalies, once developed in the tropical Pacific, tend to persist for one to two seasons because of the strong controls exerted on them by the slowly evolving upper ocean thermal field. As a result, between the observations of existing conditions in the tropical Pacific and the multi-season forecasts based on them, our climate crystal ball was sufficiently clear by mid-1997 that nations, governmental organizations, businesses, and individuals were motivated to undertake disaster preparedness, mitigation efforts, and other responses to the developing El Niño on an unprecedented scale (Changnon, 1999; Buizer et al., 2000; Glantz, 2001).

The successes in observing and forecasting climate variability during 1997–98 represent both a milestone and a challenge. This period (continued through the present) encompasses the best documented cycle of ENSO warm and cold events on record, and the forecasts of these events in many cases showed skill above that which would have been expected by chance. However, predicting El Niño and La Niña is still far from a perfect science, as attested to in assessments of forecast skill during 1997–98 (Barnston et al., 1999; Landsea and Knaff, 2000). As we learn more about the physical mechanisms underlying the ENSO cycle, we can anticipate that the skill in forecasting both warm and cold climatic conditions in the tropical Pacific will continue to improve. If so, ENSO forecast information may be used more extensively in the future for practical purposes that benefit many segments of societies worldwide.

Key points

1. The successes in observing the evolution of the 1997–98 El Niño and subsequent La Niña were in striking contrast to the situation only 15 years ago when the last major El Niño in 1982–83 was not even detected until it was nearly at its peak.
2. As the warm SST anomalies began to wane in early 1998, prediction models were largely consistent among themselves in forecasting an end to El Niño, and the development of a La Niña sometime during the second half of 1998.

3. The seeds for El Niño's demise and for the birth of La Niña were to be found below the ocean's surface.
4. Our climate crystal ball was sufficiently clear by mid-1997 to motivate disaster preparedness, mitigation efforts, and other responses to the developing El Niño conditions on an unprecedented scale.

REFERENCES

Barnston, A.G., M.H. Glantz, and Y. He, 1999: Predictive skill of statistical and dynamical climate models in SST forecasts during the 1997–98 El Niño episode and La Niña onset. *Bulletin of the American Meteorological Society*, 80, 217–43.

Buizer, J.L., J. Foster, and D. Lund, 2000: Global impacts and regional actions: Preparing for the 1997–98 El Niño. *Bulletin of the American Meteorological Society*, 81, 2121–39.

Changnon, S.A., 1999: Impacts of 1997–98 El Niño generated weather in the United States. *Bulletin of the American Meteorological Society*, 80, 1819–28.

Glantz, M.H. (ed.), 2001: *Once Burned, Twice Shy? Lessons Learned from the 1997–98 El Niño*. Tokyo, Japan: United Nations University Press.

Landsea, C.W. and J.A. Knaff, 2000: How much skill was there in forecasting the very strong 1997–98 El Niño? *Bulletin of the American Meteorological Society*, 81, 2107–19.

McPhaden, M.J., 1999: Genesis and evolution of the 1997–98 El Niño. *Science*, 283, 950–4.

McPhaden, M.J. and X. Yu, 1999: Equatorial waves and the 1997–98 El Niño. *Geophysical Research Letters*, 26, 2961–4.

McPhaden, M.J., A.J. Busalacchi, R. Cheney, J.R. Donguy, K.S. Gage, D. Halpern, M. Ji, P. Julian, G. Meyers, G.T. Mitchum, P.P. Niiler, J. Picaut, R.W. Reynolds, N. Smith, and K. Takeuchi, 1998: The Tropical Ocean-Global Atmosphere (TOGA) observing system: A decade of progress. *Journal of Geophysical Research*, 103, 14,169–240.

WMO (World Meteorological Organization), 1999: *The 1997–98 El Niño Event: Scientific and Technical Retrospective*. WMO-No. 905. Geneva, Switzerland.

Definition(s) of La Niña

James O'Brien

Cold events in the central equatorial Pacific have been referred to in several ways: one of the most popular labels has been La Niña. Other descriptions include but are not limited to the following phrases: a cold episode, El Viejo, a cold phase of ENSO, El Niño's counterpart, a cold counterpart of El Niño, the inverse of El Niño, a cold weather version of El Niño, a cold phase, El Niño's sister, anti-El Niño, seasons with cold sea surface temperatures, an anomalous cooling, the opposite of El Niño, a sisterly event, El Niño's cold water opposite, the flip side of El Niño, abnormally cold, a periodic abnormally cold sea surface current, the girl child, the other extreme of the ENSO cycle, a mature cold episode, ENSO's lesser-known twin, non-El Niño, and so forth. Each of these descriptions has already been used by one scientist or another and by the media.

In a strict sense, La Niña is an extreme cooling, not just any below-average cooling, of sea surface temperatures in the central equatorial Pacific. However, some people have suggested that for many parts of the globe La Niña has been viewed as an extreme case of normal. With an increase in attention to the La Niña phenomenon in future years, a finer distinction will likely be made by researchers between La Niña and normal conditions in the tropical Pacific. One recent example of a shift in perspective to separate La Niña from normal conditions appeared in a report on La Niña's potential impacts in Indonesia. The report noted that "during La Niña years, the onset [of the monsoon] is earlier than normal in most of the areas ... therefore La Niña has potential for advancing the planting season and an early increased harvest ..." (Kishore and Subbiah, 1998, p. 21).

Definitions are arbitrarily true. Therefore, one must make explicit his/ her definition so that others can understand what s/he is talking about. However, when several definitions of the same phenomenon are at play, it becomes difficult to compare research analyses and findings. With regard to El Niño and La Niña, various researchers, groups of researchers, and research centers prefer one definition (usually theirs) over those provided by others. As a result, they produce lists of years during which cold or warm events occurred, according to their definitions. A major problem for researchers who focus on ENSO's impacts is that these various lists do not necessarily agree. This makes it difficult to correlate cold event occurrences reliably with, for example, the yields of specific crops or with the volume of fish catches. The problem becomes more acute because there has been only a relatively small number of La Niña events (when defined as *extremely* cold sea surface temperatures in the equatorial Pacific Ocean for an extended period of time). Scientists generally agree that, since the early 1970s, there have been about twice as many El Niño as La Niña events. According to one definition, there have been four La Niña events since the early 1970s. By another more strict definition of La Niña, only two La Niñas occurred in this period.

Altering the quantitative aspects of a La Niña definition can change the number and frequency of events (up or down) that one would include in assessments of the phenomenon or its impacts. The less strict the definition, the more La Niña events can be found. As Anthony Barnston noted (see Barnston, this volume), "when the limit [or threshold set of criteria for defining a La Niña] is too strict, the events included are the strong ones, but the sample size is small. When the limit is more lenient, a larger sample can be generated, which is good for circumventing flukes but runs the risk of including La Niñas of very different strengths and flavors in the same classification." Thus, in the absence of an agreed-upon definition by scientists of La Niña, it becomes more difficult for climate impact researchers to identify, with some degree of confidence and reliability, important La Niña-related climate anomalies and their societal and environmental impacts in either tropical or extra-tropical regions.

At the La Niña Summit, it was suggested that perhaps another way to define La Niña was by its set of worldwide teleconnections and societal impacts. However, various problems with such an approach were pointed out: (1) Given the increasing reliability and credibility of forecasts, preventive or mitigative actions are increasingly being taken by countries to reduce the severity of first- and second-order impacts (e.g., in anticipation of forest or brush fires, underbrush can be cleared away). In other words, La Niña events will continue but their adverse impacts can vary as a result of societal responses to forecasts; (2) During the 1997–98 El Niño, the Indian monsoon did not fail as had been expected. Also, the Australian winter wheat crop was successful because of a few weeks of timely rains, despite meteorological drought conditions throughout much of the growing season (Dick, 1997). There will be exceptions in La Niña's impacts in some locations from one event to another; (3) During a weak La Niña, its impacts in distant locations (e.g., teleconnections) are greatly weakened and, as a result, local and regional climate conditions are more likely to be determined by local factors than by relatively small sea surface temperature anomalies in the distant equatorial Pacific (the same, of course, is true for El Niño as well).

It was also suggested at the Summit that La Niña could be defined by various "users" of information, as dictated by their needs. This may be appealing to knowledgeable users of La Niña (and El Niño) information, because they are in a better position to identify and use the basic time series (e.g., sea surface temperature, the Southern Oscillation Index (SOI), outgoing longwave radiation) that best suits their needs for various parts of the equatorial Pacific (e.g., Niño1, Niño2, Niño3, Niño3.4, Niño4, or Niño C [see Wang, this volume]). However, this approach may be of little value to potential users of such information who are less knowledgeable about the science of Pacific air-sea interactions. "Potential users" refers to those who have not yet realized the value to their decision-making processes of La Niña information, including forecasts. Converting potential users to actual users is an important, but difficult, challenge for the ENSO forecast applications and impacts communities.

For those concerned about correlating the occurrence of cold events with changes in various human activities (agricultural production, fish catches, damages associated with extreme events such as droughts, floods, hurricanes, etc.), too many definitions can be confusing to the unsuspecting public. For example, the use of any single, potentially incomplete, or unreliable list of La Niña years can produce statistically misleading results about impacts, applications, and forecast value. Although a La Niña can be weak, moderate, strong, very strong (intense), or extraordinary, the level of "strength" of a La Niña event has as yet to be agreed upon.

The global climate system changes on various time scales from seasonal, to interannual, to decadal and beyond. Today, there is considerable speculation about how the ENSO "cycle" would be affected by a human-induced global warming of the atmosphere (see Trenberth, this volume). That there have been about twice as many El Niño as La Niña events since the early 1970s has been cited by some researchers as an early sign that global warming has begun to show its effects. However, a reconstruction of the SST time series back to 1867 has indicated that, over this period of time, about an equal number of these events has occurred in the equatorial Pacific.

To some scientists a precise definition of a natural phenomenon, such as La Niña or El Niño or more generally the ENSO "cycle," has not been viewed as an important part of the problem associated with forecasting and forecast application. They argue that knowledgeable users were not likely to focus on lists of La Niña or El Niño years but, as noted earlier, were more likely to use the time series they considered to be relevant to their beliefs. Others, however, have pointed out that most forecast users were likely to rely on whatever definition of La Niña happened to be presented to them, without necessarily realizing that other definitions also exist. Thus, it would be beneficial for users to learn how to interpret the various time series related to the ENSO cycle (SSTs, SOI, OLR [outgoing longwave radiation], etc.) or at the least to realize that there are different ways to define ENSO's extremes.

There are other problems with defining El Niño and La Niña. If one uses sea surface temperature as the indicator data, one has the choice to rely on one of several "Niño boxes" for identifying a monthly anomaly. Some use Niño3 time series or JMA's (Japan Meteorological Agency's) definition region, which are the same except for the way they average data. Recently, Niño3.4 has been added to the existing "boxes." However, the periods of time over which the averages are determined may be different, leading to a different collection of dates of ENSO's extremes.

As a final note, in 1998–2000, the concept of a two-year La Niña arose, since this cold event lingered from July 1998 through June 2000. Scientists speak about medium and strong events, but not about the duration of these events. Such information, however, may be more important for those involved in the application of La Niña information to societal needs. There is much to do in order to categorize our knowledge about La Niña.

Key points

1. Until now, most articles have referred to La Niña in the context of the El Niño phenomenon. As a result, the reference point for La Niña (cold event) has been El Niño, whereas the reference point for El Niño continues to be "normal" sea surface temperatures (SSTs) in the central Pacific.
2. Because of different La Niña definitions, the reliability of statistical correlations varies between La Niña events and their worldwide linkages to anomalies in the environment and impacts on society.

3. Physical scientists appear to be less concerned about the need for a universally agreed-on definition or an agreed upon list of La Niña periods than are various users of La Niña information.
4. Should scientists consider changing the definition of what constitutes a La Niña event over time, in order to take into account decadal-scale changes in air-sea interactions and their worldwide impacts?

REFERENCES

Dick, A., 1997: $1 billion rain: Another soaking for El Niño. *The Land (Rural Weekly)*, 9 October, p. 1.

Kishore, K. and A.R. Subbiah, 1998: *La Niña 1998–99: Challenges and Opportunities for Indonesia*. Bankok, Thailand: Asian Disaster Preparedness Center (ADPC), p. 21. Online at www.adpc.ait.ac.th/ece/ir/report-2.html

Part 2

Aspects of La Niña

What constitutes "normal?"

Joseph Tribbia

An important question is "what constitutes normal in the context of the ENSO cycle?" Here, the focus is less on what constitutes the climatological normal than on what one thinks of or perceives to be normal. Four concerns are raised here for discussion purposes.

"Normal is what we expect"

Expectations about changes in the future state of sea surface temperatures (SSTs) in the tropical Pacific, and about the climate-related impacts brought about by those changes, do not always match well with the reality of what takes place. However, any actions taken that are based on the perception of expected changes in oceanic or atmospheric conditions, e.g., expectations prompted by forecasts of El Niño or La Niña conditions, will have real consequences for and impacts on society.

Over the past half century, our understanding of the relationship of the ENSO process to weather anomalies has grown along with our recognition of the cyclic nature of ENSO. Because of this knowledge, our expectations have changed. We know then that normal does not mean that the tropical ocean is in a near steady state (a statistically calculated average) most of the time. Variations in SSTs are to be expected and, in fact, are part of the normal ENSO process. However, societies often tend

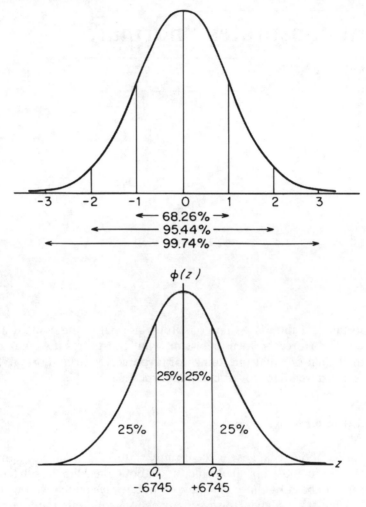

Fig. 2-1 Bell curve (after Mode, 1966).

to minimize (i.e., discount) the possible occurrence of what they consider, rightly or wrongly, to be unusual or rare events. Instead, they tend to consider only the middle section of the bell-shaped distribution curve as normal, thereby omitting consideration of the "wings" of that bell-shaped curve (Fig. 2-1). Hence, they set themselves up to be surprised when one of the ENSO extreme events does occur. There have been a number of surprises related to the ENSO cycle of warm and cold events that ought to have been expected, given the knowledge about ENSO that existed at the time.

"Normal might not be likely to occur"

Looking at SST anomalies in the tropical Pacific, one might get the impression that there is a normal (i.e., average) condition to which the temperature state of the ocean always returns. However, if one calculates the time average SSTs for a long period (say 30 years) and uses this as an estimate of "normal" SST conditions, it will be clear that the tropical Pacific is only rarely exactly normal, because SSTs are frequently in transition between warm and cold conditions. If the definition of what constitutes normal SSTs is expanded from the mathematical average temperature to a range of temperatures about the average, the probability will be higher that the SSTs are somewhere within that range. Such an expansion of the concept of normal in the context of SSTs in the central Pacific, in general, and for El Niño and La Niña, specifically, while common in science, needs to be reinforced in the user communities.

"Symmetric variations in SSTs do not mean symmetric anomalous impacts"

Typical variations around the norm in tropical Pacific SSTs do not (in general) generate normal variations in impacts at distance locations. The influence of tropical SSTs on remote meteorological conditions, such as precipitation and temperature in remote locations, is not linear. This means that the strength of their influence does not scale in a proportional way (i.e., the stronger the La Niña, the stronger the impact in a given location). Thus, the impacts, having been enhanced by local factors, can have a large magnitude, even though the SST anomaly may have been only a moderate event. Furthermore, the societal and economic impacts of variations in precipitation in a given location are also frequently nonlinear (in speaking of linear effects, one refers to effects which are both additive and proportional. So, for example, if an SST anomaly is twice as large this year when compared to a previous year, and if the air temperature response is linear, the air temperature anomaly should be twice as large. The point is that it would not scale proportionately, and if the input SST distribution is bell shaped, the response distribution will be asymmetric, i.e., skewed), making the impacts on key sectors of society, such as agriculture and fisheries, highly skewed with regard to benefits that might be accrued or losses that might be incurred.

"Extremes are part of normal"

Even under average (i.e., normal) conditions in the central equatorial Pacific, extreme meteorological events will occur around the globe. Dry spells and even droughts can occur, even when nothing anomalous or unusual is taking place with tropical Pacific SSTs. With a normal forcing (heating up) of the atmosphere, there will still be a wide range of variability in climate anomalies in both the tropics and extra-tropics around the Pacific basin. Thus, one ought to expect a certain amount of extreme climate-related activity to occur during "normal" SST periods in the tropical Pacific. This, of course, creates a problem for those seeking to attribute (that is, to link in a cause-and-effect way) a distant climate anomaly or its socio-economic impacts to specific SST conditions in the tropical Pacific at a given point in time. Extreme meteorological events can occur at any time during the full range of SSTs in the central Pacific.

It should also be noted that the short record of SST observations may not allow for a reliable linkage between SST variations associated with ENSO and climate variations. With a short climate or SST record, expectations can be distorted by using only that portion of the record that we have observed directly. In particular, short records do not capture the decadal-scale fluctuations that occur in the ocean environment. Thus, our perspective of normal SST variations and the meteorological consequences of those variations may not be reliably translated into societal impacts of importance.

Deviations from normal conditions (however those conditions are defined) can allow for the prediction of potential socio-economic impacts. However, it would be wrong to attribute every drop of rainfall that fell in a particular region in the 1997–98 period to the recent El Niño or in the 1998–2000 period to La Niña, even if there are strong teleconnections to that region. The point is that some rain would likely have fallen there anyway. Similar considerations hold for any individual storm. However, seasonal changes in precipitation can likely be attributed to an El Niño or a La Niña, being a time-averaged quantity with smaller statistical noise.

In line with a recognition of a distribution of weather and climate impacts, users of ENSO information should be taught to seek information not only on average conditions but also on the standard deviation. The public can be taught how to understand and use probabilities.

The issue of what constitutes normal is a perceptual not a scientific problem shared by individuals, economic sectors as well as societies in general. Furthermore, it was suggested that in the US, there is considerable interest in climatic conditions that are considered to be "unexpected" or abnormal. People tend to forget previous anomalous weather and climate conditions (even fairly recent ones). They are more likely to react to conditions which are perceived to be abnormal but which, in fact, are part of what they had already experienced. That is why the media report and societies take seriously such statements as "this is the biggest rainfall in six years," or "the warmest winter in ten years." People tend to discount the past.

Societies are becoming more reliant (and dependent) on large-scale computer modeling activities to provide insights (e.g., forecasts) into what might be expected to result in and from air-sea interactions in the tropical Pacific. The limited period of observations and the shortness of truly reliable time series of SST, SOI, OLR, among other indicators of change in atmospheric or oceanic conditions, gives increased importance to enhancing the reliability of coupled general circulation models.

Key points

1. Clearly, a definition is needed of what constitutes normal with respect to SSTs in the central and eastern Pacific.
2. "Normal" should not be defined simply as "average"; it should also include the extremes. In some parts of the globe (e.g., Peru, Australia) La Niña conditions represent one facet of what might be considered "normal" in terms of expected climate conditions.
3. Even under "normal" conditions, extreme meteorological events such as droughts, floods and fires can and do occur at various locations around the globe.
4. In general, typical variations around the norm in tropical SSTs do not generate normal variations in impacts at distant locations (e.g., the colder the SSTs, does not necessarily mean the bigger the impacts in a given distant location).

REFERENCE

Mode, E.B., 1966: *Elements of Probability and Statistics*. Englewood Cliffs, NJ: Prentice-Hall, Inc.

La Niña teleconnections

George Kiladis

Teleconnections can generally be viewed as "remote influences." More specifically, in the context of La Niña, a teleconnection is the influence of cold SST on regional climate regimes that are far removed from the region of La Niña forcing, which lies in the central and eastern tropical Pacific. The following paragraphs provide a brief overview of the probable physical causes and effects of teleconnections during El Niño and La Niña.

Physical scientists are researching how events in the tropical Pacific transmit a signal through the atmosphere and ocean to distant places around the globe. Since tropical convection represents the primary "heat engine" for the global circulation of the atmosphere, changes in the location of this convection alter the global circulation, which in turn are manifested as climate anomalies. For example, as convection shifts westward from the tropical Pacific and intensifies over Indonesia and the Indian Ocean during a La Niña episode, the jet stream over the Pacific is weakened, thereby affecting the downstream wave activity (i.e., high and low pressure weather systems) moving into North America. Thus the atmosphere communicates over long distances through wave dynamics, whose forcing changes as tropical convection shifts during ENSO episodes (Trenberth et al., 1998).

Below is an updated list of El Niño and La Niña years (Table 2-1).

This list of ENSO "events" was determined by an objective method developed by Kiladis and van Loon (1988). As pointed out by Trenberth

Table 2-1 El Niño and La Niña years

Warm Events: 1877 1880 1884 1888 1891 1896 1899 1902 1904 1911 1913 1918 1923 1925 1930 1932 1939 1951 1953 1957 1963 1965 1969 1972 1976 1982 1991 1997
Cold Events: 1886 1889 1892 1903 1906 1908 1916 1920 1924 1928 1931 1938 1942 1949 1954 1964 1970 1973 1975 1988 1998

(1997), there are many possible ways to define La Niña and El Niño, and any one list will differ from others depending upon the methodology used to define an event. Kiladis and van Loon stated that a year qualified as a "warm" event (El Niño) when "the SST anomaly (within 4° of the equator, from 160°W and the South American coast) was positive for at least three seasons, and was at least 0.5 °C above the mean for at least one season, while the SOI had to remain negative and below −1.0 for the same durations." Exactly opposite conditions were used to define "cold" or La Niña events. The first year in which these conditions apply was called the "Year 0" of the event, and listed in Table 2-1, although many of these episodes were of multiple-year duration.

The list of ENSO events was used by Kiladis and van Loon (1988) and Kiladis and Diaz (1989) to statistically identify likely impacts of La Niña around the globe. This was done by a method known as "compositing," or averaging the anomalous temperature and precipitation over all the La Niña (or El Niño) events on the list, regardless of their magnitudes. In this way, if one knew ahead of time whether a La Niña or El Niño was going to develop, the statistical likelihood of a particular teleconnection could be calculated. This then provides a measure of the probable impact of a La Niña event where the signals are strongest.

Almost by definition, islands in the equatorial Pacific Ocean are very strongly affected by changes in SSTs. During La Niña conditions, islands within a few degrees of latitude north and south of the equator and eastward of 165° east longitude experience drier than normal conditions because of the stabilizing effect of cold SSTs on the atmosphere (see Figure 2-2 (c)). It should be pointed out that these signals might not really be considered teleconnections in a strict sense, since they are directly influenced by the local SST, rather than by remote forcing. At the same time rainfall is enhanced compared to normal over the southwestern tropical Pacific, eastern Indian Ocean, and Indonesia. These signals represent a shift of the main "convergence zones" of heavy tropical precipitation directly caused by a change in SST forcing.

One measure of the reliability of the observed signals is based simply on the percentage of time a meteorological station had registered above or below normal temperature or precipitation, compared to the "expected" sign and magnitude of the teleconnection signal according to the composites. It has been demonstrated that the equatorial Pacific islands have very robust signals, with virtually all La Niña events drier and all El Niño events wetter than normal in the region, based on the sample shown in Table 2-1. Similarly, Indonesia has equally robust signals of opposite sign. Away from these two regions, the signals begin to become somewhat less reliable, meaning that not all ENSO events will necessarily display the "expected" anomalies in temperature or precipitation calculated for a composite event using the entire sample.

It is frequently suggested that there is some degree of linearity between El Niño and La Niña impacts over many regions, meaning that cold events in those locations produce climate anomalies of the opposite to those that tend to occur during warm events. For the most part, this appears to be true in the regions of strongest teleconnections, as is the case for equatorial Pacific Islands and in Indonesia, for example (compare Figures 2-2 and 2-3 (c)). However, the reliability of the teleconnection signals becomes less as one moves farther away from the tropical Pacific, the "center of action" of ENSO. Thus, even though a given teleconnection might still be defined as "statistically significant," and almost certainly shows a real relationship to the entire population of warm or cold events, the probability of that signal occurring during any individual event may not be very high because of the large amount of climate "noise" (random fluctuations) in the atmosphere. This is especially true outside the tropics, where large "internal climate variability" (i.e., that variability which is generated through feedbacks within the atmosphere, not forced externally by the ocean) is dominant. This differs from the tropics where climate conditions are determined by SSTs to a much larger degree. In addition, weak and moderate La Niña events might not be strong enough to generate climate anomalies in distant locations.

Figure 2-2 (c) shows a synopsis of the more robust La Niña teleconnections. These include: a tendency for wetter than normal conditions with a risk for flooding in southern Africa and the monsoon regions of India, Indonesia, and northern Australia; and drier than normal conditions, sometimes leading to drought, over eastern Africa, the western equatorial Indian Ocean, southern South America, and the southern Great Plains and southeastern portions of the United States. The great floods in Mozambique during the 1999–2000 and 2000–2001 seasons were certainly in line with the expected signal of above normal precipitation during La Niña events, although it may be that other forcings, such as the warmer than normal adjacent Indian Ocean SST, were also factors. Other effects beyond ENSO, such as the state of the SST field outside the Pacific, are frequently taken into account in seasonal forecasting schemes used by decision makers for agricultural or energy applications. As another example, many researchers have also pointed out the occurrence of drought over the southern and eastern portions of the US during the past two years as being coincident with the expected dry conditions in those regions during La Niña.

The most pronounced extra-tropical temperature signals during La Niña are seen over North America, where there is a strong tendency for colder than normal conditions over Alaska, western Canada, and the central plains of midwestern Canada and the northern United States, and warmer than normal tendencies over the southeastern United States. In general, tropical surface temperatures tend to be below normal, with robust signals even as far away from the tropical Pacific as central Africa.

In general, there is a statistically significant tendency for tropical SST anomalies of the same sign as those in the Pacific to develop in the Indian and Atlantic Oceans three to six months following the onset of both warm and cold event conditions. These remote SST anomalies are in phase with the observed tropical surface temperature anomalies, even over the tropical continents, which also lag behind the equatorial Pacific SST by the same amount of time. If one could explain the distant SST signal, one could then also account for the tropical temperature signal, since over the ocean surface the temperature of the air follows the SST very closely. The anomalous surface solar heating, because of changes in cloudiness during warm and cold events, is a probable factor (Kiladis and Diaz, 1989). For example, during La Niña, SSTs tend to become lower than normal especially over the eastern Indian Ocean and Australasia, where enhanced rainfall and cloudiness decrease the input of solar radiation into the ocean. Peter Webster has noted that the South Asian monsoon circulation may also be a likely player in governing Indian Ocean SSTs (Webster et al., 1999), through potential air-sea interaction which is independent of ENSO that might occur in that basin. Webster also suggested that an SST dipole frequently occurs in the Indian Ocean, a phenomenon characterized by an east to west oscillation of warm and cold waters that affects precipitation especially over East Africa and Indonesia (Webster et al., 1999; Saji et al., 1999). This is an important area for future research, since, when combined with the predictability related to ENSO fluctuations, it could brighten the prospects for predicting rainfall over some of the most productive food-growing regions in the Eastern Hemisphere.

Even though a given teleconnection might be defined as "statistically significant," and almost certainly related to the total population of warm or cold events, the probability of that signal occurring during any particular El Niño or La Niña may not be very high because of random fluctuations in the atmosphere. Over 25 years ago in an interesting article entitled "Teleconnections and the Siege of Time," Colin Ramage (1983) referred to the fact that many of the time series used in teleconnection analyses are of relatively short duration, and that teleconnections identified as being robust during one epoch may fail completely during a later epoch. While some of this might be attributed to the statistical variability inherent in using short samples (not many years of data), there could also be long-term changes in the climate system which could alter the response of the atmosphere to SST anomalies. Thus, forecasts based on reliable La Niña teleconnections, even those considered highly statistically significant, could fail or even reverse sign in the future because of decadal (or longer) time-scale climate variability. Nevertheless, we should note that the most statistically reliable teleconnections identified over the past century were indeed still present in the 1997–2000 El Niño/La Niña cycle (see, for example, www.pmel.noaa.gov/toga-tao/el-nino/1997.html). The identification of slow changes in the patterns of teleconnectivity associated with El Niño and La Niña will therefore require continual monitoring of future events against the background of potential natural or anthropogenically induced climate variability. In addition, linearity, or the reversal in the sign of anomalies in the same location between La Niña and El Niño, exists to some extent on the large spatial scale of teleconnection maps such as in Figures 2-2 (c) and 2-3 (c), but would likely break down in a more detailed analysis with increased spatial resolution in many regions. This would result, for example, from the influence of mountains or the coastal ocean on local climate responses to changes in the global-scale atmospheric circulation.

Because a seasonally averaged teleconnections map suggests no clear or consistent mean influence of La Niña or El Niño on other parts of the globe, this does not mean that there is no influence of ENSO in a particular area. ENSO's impact may reside in the character of the sub-seasonal variations of weather (e.g., extreme high and low temperature, first frost date, extreme precipitation event, etc.), although these problems have only recently become a focus of meteorological research (see, for example, www.cdc.noaa.gov/ENSO/).

One fruitful area for ENSO research involves looking at the variability within a season as it relates to tropical SSTs. As an example, winter temperature over the central Great Plains in the USA during both warm and cold events averages near the long-term mean, so it might appear that there is no effect on temperature because of ENSO. However, during La Niña episodes it appears that temperature fluctuates more, with an increased risk of arctic air outbreaks balanced by anomalous warmth at other times, to give "average" over a season as a whole (Wolter et al., 1999). This type of teleconnection might have a much greater impact than a more steady anomaly of the seasonal mean temperature, particularly in agricultural applications where sensitivity to extremes of temperature or precipitation might be high.

Accurate long-term forecasting of La Niña and El Niño will ultimately depend on a better understanding of higher frequency weather fluctuations, such as "westerly wind bursts" along the equatorial Pacific. These bursts, lasting only a few days, are thought to be important in setting up the ocean to a transition from either normal or La Niña conditions to an El Niño. Such a burst is believed to be responsible for triggering the 1997 El Niño (Kerr, 1999), and an "easterly burst" over the tropical eastern Pacific has been implicated in abruptly ending that event and kicking off the 1998 La Niña episode (Takayabu et al., 1999). It seems likely that the details of these types of forcings will have to be simulated in numerical models for truly reliable forecasts of El Niño and La Niña out beyond one season. However, another school of thought in the climate community believes that the ocean does not really respond decisively to these faster atmospheric fluctuations, and that if a researcher forces an ocean model with only the low frequency components (those associated with time scales of months or more) it will still accurately produce warm and cold events. Many researchers are presently issuing ENSO forecasts using both dynamical and statistical models, and over time the important factors determining an accurate prediction will become clearer as the sample sizes of these predictions becomes larger.

Key points

1. A fruitful area for ENSO research might involve going beyond simply documenting the "expected" anomalies during a particular season due to La Niña and El Niño, and looking at the variability within a season as it relates to tropical SSTs. It appears that during La Niña, temperature fluctuates more, with increased risk of arctic air outbreaks balanced by anomalous warmth at other times. This type of teleconnection might have a much greater impact than the long-term mean temperature.
2. Accurate long-term forecasting of La Niña and El Niño will ultimately depend on a better understanding of higher frequency climate fluctuations. One school of thought in the climate business believes that the ocean doesn't really care too much about the faster atmospheric fluctuations, and that if you force an ocean model with only the low frequency components, it will accurately produce warm and cold events. However, the nature of teleconnections depends on the details of, say, the SST distribution.
3. Many of the time series used in teleconnection analyses are of rela-

tively short duration. Therefore, teleconnections identified as being robust during one epoch may fail completely during a statistical fragility of using short-term samples, or there could also be long-term changes in the climate system itself which could alter the response of the atmosphere to sea surface temperature anomalies.

4. Because a seasonally averaged teleconnection map suggests no clear or consistent influence of La Niña or El Niño, it does not mean that there is no influence of ENSO in a particular area. For example, ENSO's impact may reside in the character of the sub-seasonal variations of weather (e.g., extreme high or low temperature, first frost date, extreme precipitation event, etc.), though these problems have only recently become a focus of meteorological research.

REFERENCES

Kerr, R.A., 1999: Does a globe-girdling disturbance jigger El Niño? *Science*, 285, 322–3.

Kiladis, G.N. and H.F. Diaz, 1989: Global climatic anomalies associated with extremes in the Southern Oscillation. *Journal of Climate*, 2, 1069–90.

Kiladis, G.N. and H. van Loon, 1988: The Southern Oscillation. Part VII: Meteorological anomalies over the Indian and Pacific sectors associated with the extremes of the oscillation. *Monthly Weather Review*, 116, 120–36.

Ramage, C.S., 1983: Teleconnections and the siege of time. *Journal of Climatology*, 3, 223–31.

Saji, N.H., B.N. Goswami, P.N. Vinayachandran, and T. Yamagata, 1999: A dipole mode in the tropical Indian Ocean. *Nature*, 401, 360–3.

Takayabu, Y.N., T. Iguchi, M. Kachi, A. Shibata, and H. Kanzawa, 1999: Abrupt termination of the 1997–98 El Niño in response to a Madden-Julian Oscillation. *Nature*, 402, 279–82.

Trenberth, K.E., 1997: The definition of El Niño. *Bulletin of the American Meteorological Society*, 78, 2771–7.

Trenberth, K.E., G.W. Branstator, D. Karoly, A. Kumar, N.-C. Lau, and C. Ropelewski, 1998: Progress during TOGA in understanding and modeling global teleconnections associated with tropical sea surface temperatures. *Journal of Geophysical Research*, 103, 14,291–324.

Webster, P.J., A.M. Moore, J.P. Loschnigg, and R.R. Leben, 1999: Coupled ocean-atmosphere dynamics in the Indian Ocean during 1997–98. *Nature*, 401, 356–60.

Wolter, K., R.M. Dole, C.A. Smith, 1999: Short-term climate extremes over the continental United States and ENSO. Part I: Seasonal temperatures. *Journal of Climate*, 12, 3255–72.

Climate change and the ENSO cycle: Are they linked?

Kevin E. Trenberth

There is considerable scientific and public interest in and concern about the possible impacts of global warming on El Niño events, and of El Niño on global warming. That interest continues to grow. Some studies (e.g., Trenberth and Hoar, 1996, 1997) have pointed to the unusual behavior of the tropical Pacific's sea surface temperatures (SSTs) in the first half of the 1990s, in which there were three consecutive El Niños where SSTs failed to go below average in between, and so it appears more like one continuous event (see Trenberth, 1997) from 1991 to 1995. Those studies also pointed to the pronounced climate shift about 1976 that has led to a predominance of El Niños since then, including the two biggest on record: those in 1982–83 and 1997–98. Consequently, the question has been raised about whether these changes have resulted from the influences of a human-induced global warming of the atmosphere. While a plausible case can be made, there is not yet enough evidence to identify reliably the linkages between these two processes, one natural (the ENSO cycle) and one human-induced (an increase in greenhouse gas emissions).

Different views exist over the possible connections between El Niño events (their characteristics and their impacts) and human-induced global warming. In October 1997 an "El Niño Summit" was called for and convened in California, a few months in advance of the Kyoto Conference of Parties to the Climate Convention. During the Summit, US Vice President Gore spoke of the likely influence of global warming on El Niño, suggesting that El Niño's characteristics would strengthen and its impacts would become more damaging as a result of global warming. Given the heated political disagreement over the climate change issue, the linkage between El Niño and global warming had to some extent become a politicized issue as much as a scientific one.

A Greenpeace-sponsored review of this issue, prepared for the Kyoto Conference of Parties to the Climate Convention in Kyoto, Japan in December 1997, was unable to draw a conclusion about the links between global warming and El Niño events (Karas, 1997). However, the review could not rule out such a linkage. It is highly likely that Greenpeace had hoped to be able to find conclusive scientific consensus on a positive linkage between these two processes, but was unable to do so after reviewing the scientific literature and interviewing members of the ENSO and climate change research communities. This issue is as yet to be resolved and is an important research question.

To address the global warming-ENSO linkage, the following set of questions should be addressed: How have global temperatures changed? How has El Niño changed? What is the role of El Niño in global temperature changes? Can we distinguish between naturally occurring decadal-scale variability and anthropogenic effects on global temperatures? How helpful are the paleo records? How useful are the global climate models? Even with little change in El Niño, are the effects, in terms of droughts and floods, enhanced?

There is no doubt that global mean temperatures have been increasing. According to the 1995 IPCC report "the balance of evidence suggests that there is a discernible human influence on global climate" (IPCC, 1996). Since then the evidence has become much stronger. Several years since then have increased the observational evidence and 1997 and 1998 are the warmest years on record. The decade of the 1990s is by far the warmest in the instrumental record. Moreover, new syntheses of paleoclimatic data (Mann et al., 1999) provide a much more complete context by reconstructing the Northern Hemisphere annual temperature record back 1,000 years, and the exceptional warmth of the late twentieth century becomes strongly apparent. The modeling of global climate change has improved, and increasing evidence suggests that it was in the late 1970s that the human-induced climate temperature effects emerged clearly from the noise of natural variability. This is the same time that changes in ENSO occurred. Surely this is more than a coincidence.

Of course part of this is simply that there is a mini global warming associated with and following El Niños, as was especially the case in the first half of 1998, for instance. Thus, some of the warmth of 1998 – perhaps 0.2 °C – came about because of the massive intense El Niño in 1997–98. This effect is partly the result of the cooling process that goes on in the tropical Pacific during El Niño's decay phase when some of the heat loss from the ocean goes into the atmosphere (Sun and Trenberth, 1998). But the warmth of 1999, a La Niña year, strongly indicated that the warmth was much more than the result of El Niño.

There have been questions about the reliability of SST data (Hurrell and Trenberth, 1999), and so this raises questions about how well the past record of El Niños is known. SST observations in the tropical Pacific are especially spartan prior to the 1950s. Fortunately, tropical Pacific SSTs, and especially those in the Niño regions, are strongly correlated with the Southern Oscillation Index (SOI); for relations of the SOI with temperatures and precipitation see Trenberth and Caron (2000) (www. cgd.ucar.edu/cas/papers). The SOI does provide a fairly homogeneous time series with which to explore past ENSO behavior for the period after about 1880. It is this series (Figure 2-4) that shows how unusual the post-1976 period has been.

Another approach to identify the interactions between global atmospheric temperature and ENSO events is to use computer-based climate models. In this case the verisimilitude of the simulation of the mean and variability in the tropical Pacific is critical. At present, no global climate model has been able to simulate El Niño events as realistically as desired in order to build confidence in its projections. Several models, but not all, suggest that global warming would increase warming in the eastern Pacific, so the mean climate would become a bit more El Niño-like. Some models suggest that stronger El Niño-La Niña swings may occur (e.g., Timmermann et al., 1999), even as the total climate becomes more El Niño-like. This comes about because of greater warming at the surface, and thus an enhancement of the upper ocean temperature gradients in the thermocline. So, while climate models certainly show changes with global warming, none simulate El Niño with sufficient fidelity to generate confidence in the results. How clouds might change, especially the brightness of convective clouds, is especially uncertain and can influence the outcome. The question of how El Niño may change with global warming is very much an important research topic.

The instrumental observational record is not really long enough (about 125 years for the SOI) to sample all of nature's variability, especially with regard to decades-long fluctuations. One approach to expand the length of the time series is to use proxy data, such as tree rings, the annual layers of coral, and tropical glaciers. Some reconstructions have been made using these methods, and they show El Niño variations quite nicely. However, their decadal-scale variability record is compromised by non-climatic effects such as the growth of trees or coral colonies, and biological and chemical influences. So, there is a need for multiple reconstructions using nearby cores to discern the common signal, which is likely to be from climate, from the potentially spurious component related to the

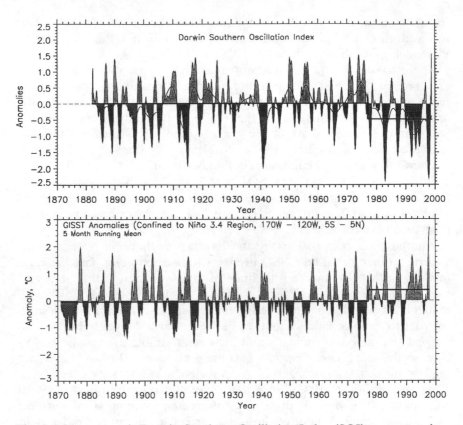

Fig. 2-4 Upper panel: Darwin Southern Oscillation Index (SOI) represented as monthly surface pressure anomalies in millibars. Data span January 1882 to December 1998. Base period climatology computed from the period January 1882 to December 1981. Lower panel: Niño3.4 region (120W–170W, 5S–5N) monthly SST anomalies in C derived from the UK Meteorological Office Global Ice and Sea Surface Temperature (GISST) data set. Data span January 1871 to December 1997. Base period climatology computed from the period January 1950 to December 1979.

individual core. Another problem with proxy data is that because there have been real climate changes in the past, such as the Little Ice Age, the environment on which ENSO operates is not constant and this too influences the variability. The result is that proxy data identify the total variability and fluctuations that occur in a changing climate, not the total variability that occurs in an unchanging climate. It is the latter that would provide the best baseline for judging how unusual the recent changes have been.

According to the Trenberth and Hoar (1996, 1997) calculations and methods of analysis, there is a very low probability that global warming is not influencing El Niño. Because ENSO is involved with the movement of heat, it is conceptually easy to see how increased heating from the build up of greenhouse gases can interfere with the ENSO cycle. Reasons for the unusual behavior of El Niño in the past 20 years may be that the "warm pool" in the western tropical Pacific is becoming larger, the recharge of heat feeding El Niño is faster and/or the heat loss from the ocean is less efficient, all of which would suggest the future likelihood of more El Niño events. With greater warming in the upper layers of the ocean, the vertical temperature gradients in the ocean could sharpen, potentially increasing the magnitude of ENSO events. Thus, we should expect big changes in El Niño's behavior although we do not yet know exactly what those changes will be or how they will affect atmospheric teleconnections worldwide.

But what about the fact that the past two years have been dominated by La Niña conditions? As noted here, the ENSO cycle is normal, and is unlikely to cease unless major perturbations occur. La Niñas are indeed still expected even if the overall climate becomes more El Niño-like. Nor should two in a row be considered out of the ordinary. But this does raise questions about whether the 1976 climate shift is permanent or temporary, and whether decadal variability might not be sending things back in another direction? However, evidence of the demise of the La Niña was strong in the northern spring of 2000, and the next El Niño appears to be looming as a pronounced warming at about 100 m depth in the western and central tropical Pacific. Time will tell.

Finally, even if ENSO does not change, the hydrological cycle is expected to be enhanced with global warming, and so the regional droughts and floods that occur with El Niño (and which tend to reverse with La Niña) are likely to become more severe.

Key points

1. The paucity of data makes establishing a reliable, long El Niño record difficult. Proxy data are not yet up to the task. The SOI is a good simple (single number) index to monitor the state of the tropical Pacific on time scales beyond interannual.
2. There is strong observational evidence (Fig. 2.4) which shows that something unusual is happening in the tropical Pacific. Models are not yet adequate to clarify what this unusual happening is or what is causing it to occur.
3. The evidence for global climate change arising from changes in atmospheric composition because of human activities continues to mount, and the human signal appears to have emerged from the background noise in the last two decades.

4. While it is reasonable to assume that global warming would affect ENSO, it is not yet known how, although hypotheses exist. While we can say that we have detected a change, we cannot do the attribution, we cannot say which part of what we are seeing is the result of global warming, although there is a very strong case that some part is.
5. Following El Niño's peak, the global atmosphere temporarily warms up. The opposite occurs following La Niña's peak. The fluctuations are at most a few tenths of a degree Celsius.
6. ENSO greatly affects the regional climate, especially in Pacific Rim countries, and so how it changes with global climate change is critical in these regions, in particular.

REFERENCES

Hurrell, J.W. and K.E. Trenberth, 1999: Global sea surface temperature analyses: multiple problems and their implications for climate analysis, modeling and re-analysis. *Bulletin of the American Meteorological Society*, 80, 2661–78.

IPCC, 1996: *Climate Change 1995: The Science of Climate Change*. J.T. Houghton, F.G. Meira Filho, B.A. Callander, N. Harris, A. Kattenberg, and K. Maskell, (eds.). Cambridge, UK: Cambridge University Press.

Karas, J., 1997: *Troubled Waters: El Niño and Climate Change*. Amsterdam, The Netherlands: Greenpeace International Report.

Mann, M.E., R.S. Bradley, and M.K. Hughes, 1999: Northern Hemisphere temperatures during the past millenium: inferences, uncertainties, and limitations. *Geophysical Research Letters*, 26, 759–62.

Sun, D.-Z. and K.E. Trenberth, 1998: Coordinated heat removal from the tropical Pacific during the 1986–87 El Niño. *Geophysical Research Letters*, 25, 2659–62.

Timmermann, A., J.M. Oberhuber, A. Bacher, M. Esch, M. Latif, and E. Roeckner, 1999: Increased El Niño frequency in a climate model forced by future greenhouse warming. *Nature*, 398, 694–6.

Trenberth, K. and T.J. Hoar, 1996: The 1990–1995 El Niño-Southern Oscillation event: Longest on record. *Geophysical Research Letters*, 23, 57–60.

Trenberth, K.E., 1997: The definition of El Niño. *Bulletin of the American Meteorological Society*, 78, 2771–7.

Trenberth, K.E. and J.M. Caron, 2000: The Southern Oscillation revisited: Sea level pressures, surface temperatures and precipitation. *Journal of Climate*, 13, 4358–65.

Trenberth, K.E. and T.J. Hoar, 1997: El Niño and climate change. *Geophysical Research Letters*, 24, 3057–60.

The symmetry issue

Martin Hoerling

The question of symmetry in the physical processes behind El Niño and La Niña events is both interesting and important. One sometimes hears the question "Are there only two states in the tropical Pacific Ocean: El Niño and non-El Niño (e.g., climatic conditions perceived to be normal)?" Or are there three states, namely El Niño and La Niña, and a so-called "normal" condition? Scientific evidence is convincing in demonstrating that the tropical Pacific Ocean undergoes an oscillation, albeit irregular, between warm El Niño conditions and cold La Niña conditions. This variability occurs relative to a climatological average, or normal ocean state. Sometimes, this normal state is little more than nature's brief respite between transitions from El Niño to La Niña. The normal state is, however, the most common occurrence for the sea surface temperatures (SSTs) in the sense that a frequency diagram of the SST anomalies shows the most likely occurrence to have near-zero anomalies. The overall distribution resembles a bell-shaped curve (or Gaussian distribution). The El Niño and La Niña events reside in the less-frequented tails of that curve.

But is that distribution a perfect bell shape, or is it somehow skewed? For example, can the sea surface temperatures become anomalously colder by as many degrees as they can become anomalously warmer? Does the atmosphere – by which we refer to the rainfall, pressure, and wind systems – respond symmetrically with respect to El Niño and La Niña? In other words, is the atmosphere's response to El Niño merely the mirror image of its response to La Niña? Does the atmosphere react twice as strongly if the ocean temperature anomalies are doubled? To the extent that answers to the above are in the affirmative, one would say that there exists a linear relation between the SST anomalies and their climatic impacts. This, of course, would greatly simplify our ability to predict the expected effect of tropical Pacific SST anomalies on climate, but the problem is somewhat more complicated than this simple linear (symmetric) view of the world.

One can address the issue of symmetry between El Niño and La Niña by using as a tool the results from several sophisticated numerical models of the atmosphere and their sensitivity to ENSO extremes. George Kiladis (in the Teleconnections chapter) had summarized 120 years of climate data, of which nearly 50 were classified either as El Niño or La Niña years. His point was that the results from such an empirical analysis would tend to yield a linear view of the atmosphere's response by default, since the data record was virtually split into two equal parts. His purpose was to study the atmosphere's sensitivity in the extreme tails of the ENSO bell-shaped distribution, since it is the extreme events that undoubtedly exert the largest impact, and for which any departures from symmetry would be critical to understand. Needless to say, the observational data offer only a few such extreme cases, though the 1997–98 and the 1982–83 Northern Hemisphere winters were during the strongest warm events in the instrumental record. The last extreme La Niña was during the winter of 1988–89 and, before then, 1973–74 (see figs. 1-11, 1-12).

Some examples from the observational data alone suggest that, if one separates out the events according to their varying intensities, one finds that the weaker events tend to behave in a linear fashion (as suggested in the overall analysis of Kiladis). It is with the larger ENSO events that one tends to see asymmetry or nonlinear behavior in atmospheric and oceanic anomalies, although those assertions were derived from an analysis of only the four cases noted in the previous paragraph (Hoerling et al., 1997). My main point was that in several parts of the globe, including North America and even the tropical Pacific, the wintertime (defined as December, January, and February) climate anomalies associated with the extreme La Niña events were weaker than their El Niño counterparts. Also, the spatial patterns were not mirror images of each other, but instead the centers of maximum anomalies over the Pacific-North American region were shifted several thousand kilometers with respect to one another. These empirical results cast some doubts over the notion of simple symmetry although, as I have noted, only a few cases did not provide us with statistically significant information. However, they suggested a course of numerical experimentation the highlights from which are discussed below.

To the media and to the public, it appears that whatever transpired in the most recent major La Niña or El Niño event can serve as the baseline as to what they should expect to happen in the next event. This is a problem of "discounting the past"; that is, people tend to weight more heavily the events that were memorable or that occurred more recently than those that occurred further back in time. Another way to look at this problem relates to the use of analogues. Researchers, and then the media, stated that the 1997–98 El Niño was like the 1982–83 El Niño. Thus, by inference one is to expect that the behavior of the tropical Pacific Ocean and its distant impacts around the globe would likely be similar to those that had occurred at that time. This is a form of "anchoring", where our perceptions and expectations are tied to a previous notable event, in this case the 1982–83 event. In future years it is likely that such anchoring will shift to the most recent extraordinary 1997–98 event instead of the 1982–83 El Niño.

With regard to the La Niña event that had been forecast for the Northern Hemisphere winter 1998–99, researchers and the media have apparently anchored the public's memories to the La Niña that occurred in 1988–89. There is also the underlying notion of linearity, and so we are likewise witnessing an anchoring to the 1997–98 El Niño event, but with the expectation that impacts will assume the opposite sign.

In both cases, anchoring can produce misleading results. There have been relatively few La Niña events in the past few decades and the impacts of the few events that we have observed are not necessarily reliable indicators of what might be the meteorological impacts that accompany a future La Niña event. Thus, anchoring can yield inappropriate societal responses to La Niña.

Results from recent general circulation modeling experiments related to the issue of ENSO symmetry focused on the strong El Niño event of 1982–83, and the major cold event of 1973–74. In each case, the observed tropical Pacific sea surface temperatures were specified as a global boundary condition, and the atmospheric model was run in order to study climate sensitivity. In order to overcome the problem of statistical sampling (e.g., too few cases) which plagues observational work, the experiments were repeated 10 times using the same SSTs, but each run began from a different initial atmospheric condition. A unique aspect to the study was the execution of a parallel set of model runs. In these, the signs of the 1982–83 and 1973–74 SST anomalies were reversed in the tropical Pacific and, once again, 10 simulations of the atmospheric model were conducted for each. Thus, a perfect mirror image of the SST anomalies was created for the two ENSO extremes, and a precise analysis of atmospheric symmetry was undertaken within the idealized scenario that the SST anomalies were themselves perfectly symmetrical in sign (Hoerling et al., 2001).

Figure 2-5 (c) illustrates the response of wintertime precipitation over North America during strong El Niño conditions (top) and equally strong La Niña conditions (bottom). Despite the exact symmetry in equatorial Pacific SST anomalies used in these climate simulations, there is a lack of symmetry in the rainfall responses, especially along the Pacific west coast. This is true for both the pattern of rainfall anomalies along the west coast, and also the intensity of those anomalies. It is readily apparent that the strong El Niño forcing induces a larger disruption of rainfall over the west coast than does the equally strong La Niña forcing.

The results of this model-based experiment tended to confirm the impression gathered from the observational analysis. In particular, the atmospheric response to strong El Niño forcing is greater than it is to equally strong La Niña forcing, and their respective spatial patterns of atmospheric responses are shifted.

Several physical factors that could account for this asymmetric behavior are as follows:

1. Under normal sea surface temperature conditions during the Northern Hemisphere winter, extreme cold water exists in the central and eastern equatorial Pacific Ocean, and the pool of warm surface water (i.e. the warm pool) is confined to the western Pacific. Owing to surface warmth in the western Pacific, the surface trade winds tend to converge there and fuel an abundance of convective activity, whereas the cold tongue area is a comparative dry zone. During an El Niño, warm water covers a large expanse of the equatorial Pacific, and during an extreme event such as 1982–83, enhanced convective activity (as measured by outgoing longwave radiation [OLR]) occurs over the central and eastern Pacific. This has a major impact on the location of storm tracks across the North Pacific and North America which are central to determining temperature and rainfall regimes during winter. During La Niña, an anomalous cooling of the cold tongue region only achieves a modest further reduction of rainfall in the already arid zone of the eastern Pacific. Further cooling during extreme events yields no further effect on the local convective activity (which, by then, has already been virtually shut down). Thus, there is an obvious saturation effect of the climate response to La Niña which does not occur in response to El Niño.

2. There is a remote influence of these disruptions in the equatorial Pacific rainfall regimes that is apparent in the middle troposphere, for example the flow of air on the 500-mb pressure surface. There is roughly a linear increase in amplitude of the 500-mb response over the Pacific North American area for successively larger-amplitude positive (warm) sea surface temperature anomalies, consistently low pressure over the central North Pacific, and North American high pressure.

Evidence exists for a modest eastward shift of this wave pattern for increasingly larger positive SST anomalies. Thus, for stronger El Niño events the North Pacific low tended to be positioned closer to North America's Pacific west coast.

On the other hand, stronger La Niña SST forcing did not lead to a further amplification of the 500-mb response. There is an initial linear response, and then the relationship becomes insensitive, which is consistent with the saturation of the tropical rainfall response.

The climate system is much more deterministic with respect to tropical Pacific rainfall: a single El Niño event produces a reliably detectable signal across the tropical Pacific. Over North America, however, one needs to witness several El Niño events in order to get a statistical sampling of the climate signal related to the SST forcing, and to be able to distinguish that from the range of climate states that occur naturally in the absence of El Niño conditions. Many areas of the middle and high latitudes were shown to have little or no ENSO teleconnection signal.

The influence of sea surface temperature anomalies along the East African coast and in the vicinity of the maritime continent (Indonesia) may be at least of local relevance. These have some relation to the typical life cycle of ENSO itself, although not all SST variations owe their origins to ENSO. SST changes in the Indian and Atlantic oceans, and convection processes in the maritime region of Indonesia and Malaysia have been less well studied than those in the tropical Pacific (Hoerling et al., 2001).

During the recent La Niña event that developed in late spring of 1998 and continued through 1999, very warm SST anomalies were observed in the vicinity of the maritime continent. These achieved unprecedented amplitudes when viewed in the context of the modern instrumental record. Furthermore, their occurrence seems not directly related to the ENSO cycle, but instead appears tied to a longer time-scale variation of SSTs in that region. The local warmth of the ocean has contributed to prolonged wet conditions from mid-1998 through spring 2000 in the region of Malaysia and Indonesia, well in excess of the impact expected from La Niña's remote influence.

While the computer models are good at simulating the enhancement of rain in the eastern Pacific during El Niño, they are less skillful in simulating the impact on the rainfall systems over the maritime continent. There is a need for improvement in understanding the role of SST changes in the western Pacific, as it is unclear to what extent seasonal rainfall anomalies in the west Pacific are forced by local SSTs, or are remotely driven by east Pacific SST anomalies.

The activity, strength and location of the ITCZ (Inter-Tropical Convergence Zone) and SPCZ (South Pacific Convergence Zone) have important influences on global climate and, therefore, on the societal impacts of La Niña and El Niño. The ITCZ was more active during the cold event of 1988, and it remains to be determined what, if any, effect that may have had on the US Midwest drought in the spring and summer of 1988. Researchers have also referred to the PNA (Pacific North American) circulation pattern that consists of alternating high and low pressure arching across North America, and that pattern's sensitivity to ENSO. The PNA pattern is an important determinant of US weather and climate, though its sources of excitation are more varied that just El Niño.

Key points

1. The climate impacts of strong La Niña events are not simple mirror images of the impact of strong El Niño events.
2. The climate response tends to behave symmetrically with respect to the sign of the equatorial Pacific sea surface temperature forcing when that forcing is weak, i.e., the response associated with the most often observed "garden variety" ENSO events.
3. The spatial pattern of the climate impacts associated with extreme La Niña events is different from the pattern during extreme El Niño events. In particular, the centers of strongest impact over the Pacific-North American region were shifted several thousand kilometers with respect to one another.
4. The reliability (and, hence, the predictability) of North American climate impacts during extreme La Niña events is lower than it is for extreme El Niño events.
5. The Northern Hemisphere wintertime (December, January, February) climate anomalies associated with extreme La Niña events are weaker than their El Niño counterparts.

REFERENCES

Hoerling, M.P. and A. Kumar, 2000: Understanding and predicting extratropical teleconnections related to ENSO. In H. Diaz and V. Markgraf (eds.), *El Niño and the Southern Oscillation: Multiscale Variability, Global and Regional Impacts*. Cambridge, UK: Cambridge University Press, 57–88.
Hoerling, M.P., A. Kumar, and T.-Y. Xu, 2001: Robustness of the nonlinear atmospheric response to opposite phases of ENSO. *Journal of Climate*, 14, 1277–93.
Hoerling, M.P., A. Kumar, and M. Zhong, 1997: El Niño, La Niña, and the nonlinearity of their teleconnections. *Journal of Climate*, 10, 1769–86.

Attribution of societal and environmental impacts to specific La Niña and El Niño events

Gerald Meehl

Just about everything that happened between September 1997 and May 1998 that was either unwanted or unexpected was blamed directly or indirectly on El Niño. The ice storm in the northeast of North America in January 1998, the killer tornadoes in Florida in February 1998 and the scores of human deaths that ensued, the flooding in southern California, and the drought in Texas were linked by one observer or another to the 1997–98 El Niño. However, while few challenge the linkage between seasonal anomalies in climate in these regions and extremes of the ENSO cycle, many of the claims that the impacts of specific anomalous weather events can be attributed to a particular El Niño or La Niña must be challenged.

The costs of the impacts of the major drought in the US Midwest in 1988 have been estimated at about $40 billion. They were attributed to the 1988–89 La Niña without much question. The 1988 Midwest drought-La Niña connection, however, has recently been challenged because researchers have identified other plausible climate scenarios that could have produced droughts of similar magnitude in the region.

There have been several efforts by numerous organizations (NOAA, WMO, IDNDR (now ISDR), International Red Cross, CARE, etc.) to calculate the "cost" of the 1997–98 El Niño event (e.g., Sponberg, 1999). A recent study (Changnon, 1999) undertook such an assessment specifically for the United States. However, the results of such assessments must be closely scrutinized, because in order to produce credible results,

63

a rigorous method is required in order to discriminate between impacts clearly associated with El Niño, those that are not, and how much of the cost can be attributed to El Niño over and above the baseline cost of climate impacts in a given year. Only the use of a rigorous method would give credibility to the dollar numbers generated about the costs and benefits of an El Niño or La Niña.

In fact, there are two major issues on ENSO attributions that must be addressed. The first type of attribution is to determine the physical mechanisms in the air-sea interaction that cause La Niña events. This involves comprehensive research of the varied mechanisms of Pacific ocean-atmosphere coupling. Those mechanisms are captured to varying degrees in the coupled climate models that have been used to forecast El Niño/La Niña events.

The focus here is on a second type of attribution: the type of climate anomalies or impacts that could be attributed to La Niña. Here we separate the manifestations of occurrence of El Niño/La Niña from the climatic impacts of individual events. For the former type, a number of studies have documented low frequency, decadal time-scale variability in the tropical Pacific that could affect the occurrence of El Niño and La Niña impacts. Thus, the record-setting El Niño of 1997–98 could be characterized as having a contribution from (1) coupled processes on the interannual time scale, (2) a warm phase in the decadal-scale oscillation, and (3) a longer term warming trend in the eastern Pacific (Lau and Wang, 1999). These effects, superimposed on one another, could combine to produce the extraordinary El Niño that occurred in 1997–98. Similarly, these factors could affect the occurrence of individual La Niña events. We could think of attributing El Niño/La Niña occurrences to a combination of processes that take place on different time scales. This implies that there should be an enhanced public understanding of shifting probabilities of impacts from climate fluctuations on different time scales.

In the latter category – that is, how we can attribute seasonal mean climate anomalies to individual El Niño or La Niña events – the concept was developed of a seasonal mean anomaly as a collection of meteorological events. The quantification of this concept involves a shift in the probability distribution of certain conditions in a given season in a certain region. For example, during a La Niña, there is a greater chance for dry conditions over the southeastern US during the Northern Hemisphere winter than during non-La Niña episodes.

There is a strong, almost irresistible temptation to attribute individual weather and weather-related events to El Niño or La Niña. Although this runs counter to the concept noted above about a shift in probabilities for seasonal climate anomalies, a forecast study has been performed to see if individual meteorological events could be attributed to the El Niño of

1997–98 (Barsugli et al., 1999). Results from that study indicated, for example, that the 1998 ice storm in the Northeast United States was probably intensified from atmospheric forcing from the tropical Pacific, and that California's rains in the spring of 1998 could probably be attributed to forcing from the tropical Pacific. However, the cause of the Denver (Colorado) blizzard of October 1997 was inconclusive and could not be attributed to El Niño with confidence (Barsugli et al., 1999).

Glantz has commented on the chain of attributions linked to the ice storm in the northeastern part of North America in January 1998. Figure 2-6 was used to ask Summit participants to decide when, in their individual view, the attribution of impacts to El Niño should no longer be considered reliable.

In sum, there is agreement that the best way to view attribution was to think of the forcing from the tropical Pacific as having the effect of shifting the probabilities of collections of meteorological events over the course of a season toward wetter or drier, warmer or colder conditions. These effects are particularly attributable in certain seasons in certain regions (identified from previous events) as susceptible to being affected during either El Niño or La Niña events (e.g., as a result of a shift toward wetter or drier than normal conditions).

It is *very* important that attributions to La Niña events be reliable, because actions taken by societies will likely be taken based on the belief that La Niña events tend to spark certain kinds of adverse conditions, such as Midwest drought conditions in the summer of 1988. If such an attribution proves to have been incorrect, societies will have wasted scarce resources to respond to unlikely adverse events. Therefore, perhaps a better strategy is to promote a public understanding of shifting probabilities of impacts from climate fluctuations, including La Niña, that occur on different time scales.

Key points

There are two types of attribution: (1) "To what do we attribute the actual cause(s) of La Niña?" and (2) "What are the types of distant (teleconnected) climate anomalies or societal impacts we can attribute to La Niña?" Here we have focused on the second type.

1. The best way to view attribution is to think of changes in SSTs in the tropical Pacific as having the effect of shifting probabilities of the sum total of teleconnected meteorological events over the course of a season toward wetter or drier, warmer or colder conditions.

2. The 1988 Midwest drought/La Niña connection is now being re-evaluated, as researchers are developing other plausible climate

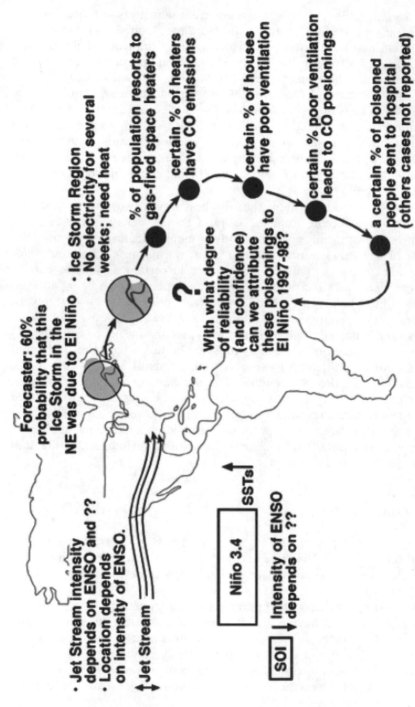

Fig. 2-6 Problem of attribution of impacts to ENSO from the January 1998 ice storm in New England (Glantz, 2001).

scenarios in addition to global warming that could produce a drought of similar magnitude in the region. This is important to note because many people tended to use the 1988–89 La Niña as a guide to their responses in the US Midwest to the forecast of the possible onset of a moderate to strong La Niña in late 1998 and early 1999. Although a major drought had been expected in this region, one did not develop there.

REFERENCES

Barsugli, J.J., J.S. Whitaker, A.F. Loughe, P.D. Sardeshmukh, and Z. Toth, 1999: The effect of the 1997–98 El Niño on individual large-scale weather events. *Bulletin of the American Meteorological Society*, 80(7), 1399–411.

Changnon, S.A., 1999: Impacts of 1997–98 El Niño-generated weather in the United States. *Bulletin of the American Meteorological Society*, 80(9), 1819–27.

Glantz, M.H. (Ed.), 2001: *Once Burned, Twice Shy? Lessons Learned from the 1997–98 El Niño*. Tokyo, Japan: United Nations University Press.

Lau, K.-M. and H. Wang, 1999: Interannual, decadal to interdecadal and global warming signals in sea surface temperatures during 1955–1997. *Journal of Climate*, 12, 1257–67.

Sponberg, K., 1999: Weathering a storm of global statistics. *Nature*, 400, 13.

Forecasting the onset of a La Niña episode in 1998

Nicholas E. Graham

The 1998–99 La Niña episode, which in fact persisted through 2000, is notable on several accounts. First, as is clear in Figure 2-7, it followed directly after the record-setting 1997–98 El Niño episode, thereby reversing climate anomalies that had been affecting many regions around the globe. Second (see Figure 2-7), this La Niña episode produced the coolest eastern Pacific sea surface temperatures (SSTs) since 1988–89, thereby ending a protracted period of relatively warm conditions. Third, the onset of the event was exceptionally rapid with much of the change from El Niño-like conditions to La Niña-like conditions occurring over just a few months. The dramatic rapidity of this change is clear in Figure 2-8, showing SSTs from the TOGA-TAO array for April and June of 1998.

Given the character and rapidity of the onset of the 1998–99 La Niña, it is of interest to consider how well various coupled models forecast these changes. To do this we will consider four coupled models: the US National Centers for Environmental Prediction (NCEP) (Ji et al., 1998; cf. www.emc.ncep.noaa.gov/research/cmb/sst_forecast), The Center for Ocean-Land-Atmosphere Studies (COLA) (Kirtman et al., 1996), Scripps Institution of Oceanography (SIO) (Barnett et al., 1993; Pierce, 1996) and the European Centre for Medium Range Weather Forecasts (ECMWF) (Stockdale et al., 1998a,b). The NCEP and COLA models each consist of an atmospheric general circulation model (AGCM) coupled to a basin-scale ocean general circulation model (OGCM) covering the tropical and sub-tropical Pacific. To make forecasts the NCEP ocean model is

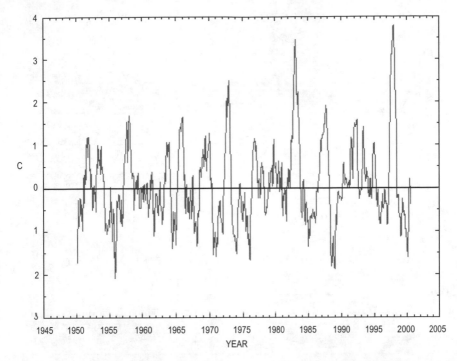

Fig. 2-7 Niño3 sea surface temperature 1945–2001 (NOAA CPC).

initialized by forcing the model with analysis winds with the concurrent assimilation of surface and subsurface temperature observations. Also, in this version of the NCEP model surface heat fluxes were handled by a statistical method called Newtonian damping. The version of the COLA model described here uses an iterative wind stress assimilation scheme to initialize the ocean model. The SIO model is a "hybrid coupled model" (HCM) composed of a basin-scale full-physics ocean model coupled to a statistical atmosphere which responds to the SSTs produced by the ocean model. Surface heat fluxes are handled through Newtonian damping. The model is initialized by calculating simulated wind stress using the statistical atmosphere driven with observed SSTs and using those winds to drive the ocean model. The ECMWF model is composed of global atmospheric and ocean models and, like the NCEP model, the ocean model is initialized through assimilation of analysis winds and data from an oceanic analysis system. Surface heat fluxes are handled in a realistic manner through surface, radiation, and boundary layer components. In the time frame under discussion, this model was unique among operational coupled models in being fully global and in producing ensemble forecasts (out to six months daily) and in its global domain.

TAO/TRITON Monthly Mean SST (°C) and Winds (m s^{-1})

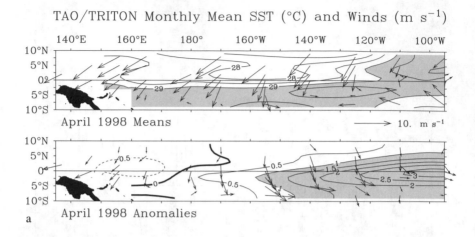

April 1998 Means

April 1998 Anomalies

a

TAO/TRITON Monthly Mean SST (°C) and Winds (m s^{-1})

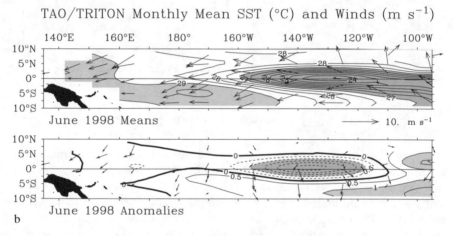

June 1998 Means

June 1998 Anomalies

b

Fig. 2-8 Sea surface temperatures from TOGA-TAO array for (a) April and (b) June 1998.

Figure 2-9 gives an idea of how well the NCEP model performed at forecasting the *onset* of the 1998–99 La Niña. In the forecasts issued from November 1997 through March 1998, the initial cooling during early 1998 was forecast relatively well. However, the predicted SSTs during the second half of 1998 were consistently too warm by 1 to 1.5 °C. The forecasts do not consistently suggest the rapid cooling observed during late spring, although the forecast issued in February was quite close with regard to that feature.

The COLA model forecasts issued from March, April, and May 1998 did predict a rapid transition to La Niña conditions (particularly the

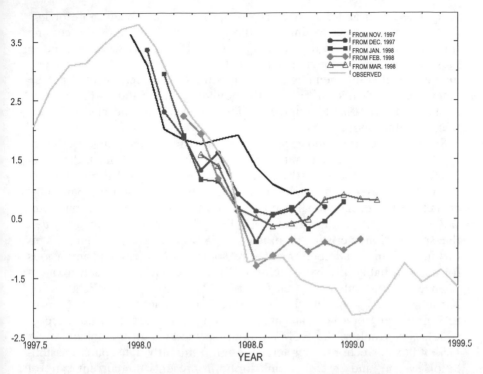

Fig. 2-9 NCEP CMB coupled model forecasts, November 1997, December 1997, January 1998, February 1998, March 1998, and observed.

latter two), although the SSTs forecast through late 1998 and early 1999 were cooler than observed. The HCM model also forecast a relatively rapid transition to La Niña conditions, but also predicted SSTs slightly cooler than observed during late 1998. Finally, the ECMWF model performed remarkably well regarding the onset of the La Niña. The forecasts hinted at the possibility of a rapid cooling during May as early as December 1997, although the forecasts for this time show a great deal of scatter. Later forecasts issued in January through April 1998 were excellent, with each suggesting very rapid cooling during May. The forecasts from February onward were also correct regarding the SSTs during the latter half of 1998.

There is no doubt that in comparison with the forecasts of the onset of the 1997–98 El Niño, forecasts of the onset of the 1998–99 La Niña event were quite successful. Probably the most serious error was the NCEP model's substantial under-prediction of the rapidity and magnitude of cooling during mid- and late 1999. These errors (up to approximately 2 °C) were sufficiently great to affect some climate outlooks issued at the time. The HCM and COLA forecasts appear to have been more or less accurate with respect to timing, although the magnitude of the cooling was over-predicted. At the same time, the initialization schemes used in these models are rudimentary and in the case of the HCM model one of the model components is extremely simple, so precision may not be among its predictive attributes. Certainly, in this instance the performance of the ECMWF coupled model was excellent and clearly better than the other models.

Successes, failures, and levels of performance in the range between, have been a time-varying characteristic of coupled models from their first applications to forecasting the future state of the tropical Pacific Ocean. Thorough analysis and insight are required to understand the particular characteristics of individual coupled models. This is true when one considers indices such as those provided by Niño3, but even more so when the spatial characteristics of predicted SST fields are considered. Unfortunately, these admonishments are sometimes overlooked in various forums (probably in this article as well), in the end to the detriment of science and the public (one useful article comparing models is given by Kirtman et al., 2000). One should also realize that accurate predictions of SSTs in other tropical oceans are not yet a reality (and may not be possible). Nevertheless, the ECMWF coupled model with its global extent, lack of flux corrections, complete (though frequently data sparse) assimilation system, and ensemble philosophy, represents the current state of the art. If its success in this instance was not happenstance, the future will see further improvements in the prediction of tropical SSTs as other new systems come on line.

REFERENCES

Barnett, T.P., M. Latif, N. Graham, M. Flugel, S. Pazan, and W. White, 1993: ENSO and ENSO-related predictability: Part 1 – Prediction of equatorial Pacific sea surface temperatures with a hybrid coupled ocean-atmosphere model. *Journal of Climate*, 6, 1545–66.

Ji, M., D.W. Behringer, and A. Leetmaa, 1998: An Improved Coupled Model for ENSO Prediction and Implications for Ocean Initialization. Part II: The coupled model. *Monthly Weather Review*, 126, 1022–34.

Kirtman, B.P. and E.K. Schneider, 1996: Model based estimates of equatorial Pacific wind stress, *Journal of Climate*, 9, 1077–91.

Kirtman, B.P., J. Shukla, M. Blamaseda, N. Graham, C. Penland, Y. Xue, and S. Zebiak, 2000: The current status of ENSO forecast skill. *A Report to the Climate Variability and Predictability (CLIVAR) Numerical Experimentation Group* (NEG), www.clivar.org/publications/wg_reports/wgsip/nino3/report.htm, October.

Pierce, D.W., 1996: *The Hybrid Coupled Model, Version 3: Technical Notes.* Scripps Institution of Oceanography Reference Series No. 96-27, August.

Stockdale, T.N., A.J., Busalacchi, D.E., Harrison, R., Seager, 1998a: Ocean modeling for ENSO, *Journal of Geophysical Research*, 103, 14,325–55.

Stockdale, T.N., D.L.T., Anderson, J.O.S., Alves, M.A., Balmaseda, 1998b: Global seasonal rainfall forecasts using a coupled ocean-atmosphere model, *Nature*, 392, 370–3.

The identification of differences in forecasting El Niño and La Niña

Stephen Zebiak

From the perspectives of dynamical or statistical models, the prerequisites for capturing the physics of both extreme cold and warm events are fundamentally the same. After extensive research over the past two decades, we have come to think of ENSO warm and cold events as different phases of a single process, an oscillation, and not as a sequencing of random events (e.g., Zebiak and Cane, 1987; Battisti and Sarachik, 1995; Neelin et al., 1998). In the context of an oscillation, the same physics describes the whole cycle, including both extremes.

However, as suggested in the index for the Niño3 region of the Pacific Ocean (Figure 2-10), one can see that the detailed characteristics of warm and cold states do have some systematic differences. For example, the cold states do not depart from average to the same extent that warm states do. Also, the frequency of warm and cold events appears to vary differently over time. Some decades, such as the 1930s, were relatively inactive periods with regard to extreme ENSO events, while others have been dominated either by cold events (earlier in the century) or by warm events (in recent decades).

There are physical differences between relatively warm and cold events relating to thermocline depth changes, the spatial expansion of convection, and the movement of the zone of atmospheric convection. In a cold state (but not necessarily a La Niña), the thermocline in the eastern equatorial Pacific is shallow to start with. It does not take much to bring it relatively closer to the surface. For the models, the turbulent mixing

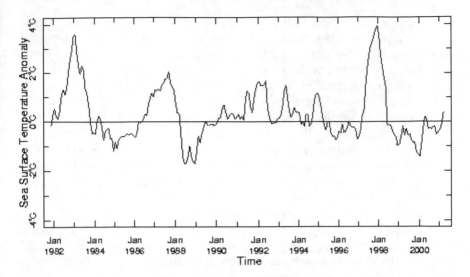

Fig. 2-10 Niño3 index.

processes in the surface layer (the zone above the thermocline) are somewhat different in El Niño as opposed to La Niña. This difference taxes the
ocean models in different ways. During strong cold events, the parameterizations are particularly taxed. The atmosphere parameterizations are
also taxed differently in warm vs cold extremes of ENSO. The response in
the extratropics may vary according to the longitude of strongest convection, which fluctuates with the phase of ENSO. From early prediction
work, there was a perception that forecasting La Niña has been more
troublesome than forecasting El Niño. More recent versions of the early
forecast systems, and newer, more comprehensive forecast models, have
not substantiated this view. Although some of the detailed physics at work
in the two extremes are different, the predictabilities of El Niño and La
Niña appear not to be much different.

It is notable that the Lamont-Doherty Earth Observatory (LDEO)
model missed the forecast of the 1997–98 El Niño – the first such miss
since its introduction in the mid-1980s. The same model (LDEO3) performed much better in a hindcast of this event, with the use of real
(observed) ocean temperatures (Chen et al., 1998). Although questions
still remain as to the apparently unusual sensitivity of the forecasts in
this event, the results underscored the value and importance of ocean
observational data to ENSO modeling activities. The ECMWF model
projections for the 1997–98 event were encouraging, as were those from
NCEP (see Graham, this volume).

Four factors can be highlighted that in general tend to limit forecast skill: model flaws, flaws in the way that data are used (e.g., assimilation and initialization), gaps in the observing system, and the inherent limits of predictability. At this point, the evidence suggests that inherent limits of predictability are not the primary limitation. There is room for further improvements, particularly in forecast initialization and data assimilation.

Forecasts are fundamentally probabilistic statements. If presented as deterministic, some will always be erroneous. Egregious forecast errors are never welcome, but they can have value for researchers in stimulating support for additional research. Errors remind scientists as well as the public that the El Niño puzzle is not fully solved. Members of the ENSO forecast community should not shy away from making their projections and predictions public, and the public and policy makers should be made aware that there will be forecast misses as well as successes.

Does the terminology used to describe El Niño events – weak, moderate, strong, very strong, extraordinary – apply to La Niña events as well? Because the range between minimum and maximum temperature anomalies for cold events is narrower than it is for warm events, it would be possible but more difficult to apply these terms to a cold event. There are fewer meaningful strength categories.

Some researchers believe that there is no "normal" state as such, and that there are only two SST states: cold and warm. Some researchers contend that what we now consider normal may in fact be only a weak state of La Niña. Perhaps there is a need to reconsider how these warm and cold states in the tropical Pacific are defined. SSTs may not the best way to characterize changes in the tropical Pacific. The conjunction of SSTs and other indicators such as sea level pressure or the positioning of convection might provide a better way to characterize the whole system in the tropical Pacific.

Key points

1. The predictability of La Niña and of El Niño are not much different.
2. The factors that limit forecasting skill for cold and warm events include but are not limited to the following: model flaws, flaws in the way data are used (e.g., in assimilation, initialization), spatial and temporal gaps in the observations and in the observing system, and the inherent limits to predictability of the atmosphere.
3. Perhaps the natural states of the tropical Pacific Ocean are warm and cold states, and there is no such thing as a "normal" state. Normal, then, would be no more than a weak state of La Niña.
4. ENSO forecasts issued to the public serve several useful purposes: they draw attention to the phenomenon, they draw attention to the needs

of the ENSO research community, and they inspire future generations to engage in research aimed at solving the ENSO "puzzle."

REFERENCES

Battisti, D.S. and E.S. Sarachik, 1995: Understanding and predicting ENSO. *Review of Geophysics, Supplement*, 1367–76.
Chen, D., M.A. Cane, S.E. Zebiak, and A. Kaplan, 1998: The impact of sea level data assimilation on the Lamont model prediction of the 1997–98 El Niño. *Geophysical Research Letters*, 25, 2837–40.
Neelin, J.D., D.S. Battisti, A.C. Hirst, F.-F. Jin, Y. Wakata, T. Yamagata, and S. Zebiak, 1998: ENSO theory. *Journal of Geophysical Research*, 103, 14,261–90.
Zebiak, S.E. and M.A. Cane, 1987: A model El Niño Southern Oscillation. *Monthly Weather Review*, 115, 2762–278.

Monitoring La Niña

Antonio J. Busalacchi

Historically, El Niño researchers have used regions in the tropical Pacific referred to as Niño3 and, more recently, Niño3.4 (Figure 2-11) as a means of tracking and diagnosing the evolution of ENSO variability. In particular, indices of SSTs are generated and have proven useful for describing the onset and evolution of an El Niño event. An important question is whether these historical indices are appropriate for describing and understanding the character of a La Niña as well. Are there gaps in the monitoring system, when viewed from a La Niña perspective? Stated another way, is what we believe to be an appropriate monitoring system for El Niño equally appropriate for monitoring La Niña?

One could argue that the TOGA-TAO array was originally set up as an equatorial wave guide array (centered on the equator) to monitor the development of an El Niño (warm event), i.e, to monitor wind and sub-surface thermal changes within the equatorial wave guide (where the equatorial ocean waves have a large amplitude). The construction of this system began in 1985, with a major influence on its development coming from the occurrence of the 1982–83 El Niño, then labeled as the El Niño of the Century. This monitoring system was fully developed by the mid-1990s.

The life cycle of the 1997–98 El Niño event (its onset, growth, mature, decay phases) was well captured by the TAO array which is situated between 8° North and 8° South of the equator. The onset phase of El Niño was well captured by the TAO array, which showed a change in the

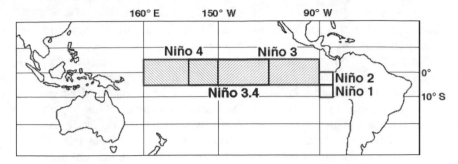

Fig. 2-11 Niño regions in the equatorial Pacific Ocean.

depth of the thermocline. Thus, changes in temperature at depth and changes in winds at the surface were monitored in a timely way: in the El Niño's mature phase, sea surface temperatures were on average peaking at 4–5 °C, with 10–12 °C anomalies at depth in the ocean. Large anomalies were observed at the surface as a result of the relaxation of the trade winds (Figure 2-12 (*c*)).

With respect to the monitoring of the onset of La Niña in 1998, the TAO subsurface temperature data in February 1998 showed a 2–3 °C cold anomaly. The TOPEX-Poseidon altimeter was in orbit at the time, monitoring changes in sea level height across the tropical Pacific. The satellite altimeter data served to complement the in situ observations of the TAO array that are restricted to within 8°N–8°S. During La Niña (cold) events, in response to stronger than normal equatorial easterlies, the sea level is higher in the western Pacific than in the eastern Pacific and along the west coast of South America. During El Niño, the sea level height in the western Pacific drops, and it increases in the eastern Pacific.

In March 1998 the development and eastward progression of cold anomalies of 5–6 °C (at depth) was apparent in the observational record. Altimetric measurements also showed an upwelling signal reflected as a drop in sea level propagating eastward along the equator. However, a very significant depression in sea level and elevation of the thermocline was observed *outside* the bounds of the TAO array at 10–15°S in the southwestern tropical Pacific Ocean. Anomalous changes in satellite observations of the surface wind stress curl field (i.e., the rotational component of the surface wind stress), were also shown to be coincident with this significant decrease in sea level (Picaut et al., 2001). These changes, poleward of the TAO domain, also raise the question regarding whether meridional (north-south) winds are as important or more important than zonal (east-west) winds? For the onset of El Niño, it is the zonal component of the wind field along the equator that is important. However, for the onset of the 1998–2000 La Niña, it was evident that there was a significant signal in the wind stress curl field off the equator. Thus, both the zonal and meridional components of the wind field are important.

During the El Niño event, there was a suppression of the nutrient-rich upwelled deep water near the Galapagos Islands. Between January and February 1998, a dramatic bloom in phytoplankton occurred across the ocean, as viewed by the SeaWiFS satellite instrument. This bloom was not along the equator but occurred a few degrees to the north. Such a dramatic change in the marine ecosystem raised questions regarding what was happening. In fact, La Niña was coming and this biological indicator was observed in advance of other more "traditional" indicators from the physical climate system. As a result of the ocean nutrients already being in a depleted state, apparently only a small amount of upwelling was needed to resupply nutrients to the euphotic (sunlit) zone and induce a chlorophyll bloom. In comparison, it took several more months of upwelling to erode the warm SST anomaly and induce a changeover to cool SST conditions normally used to indicate the presence of La Niña. For example, in April and May 1998, the upwelling signal of La Niña continued along the equator, reflected as a depression of TOPEX/Poseidon-monitored sea level and an inference of a shoaling thermocline (Murtugudde et al., 1999a).

In summary, for the onset and progression of the 1997–98 El Niño, the TAO array encompassed the field of action, as it were. However, for the changes and trend toward a La Niña, a large fraction of the signal was *outside* the 8°N and 8°S latitude bounds of the TAO array and was not captured. Furthermore, anomalous changes in sea surface temperatures in both the Indian and Atlantic Oceans also influenced the teleconnections usually associated with El Niño (Murtugudde et al., 1999b). This underscored the need for enhanced monitoring beyond the equatorial Pacific. Satellite observations contributed to this enhanced monitoring.

As for the future, there is also a need for improving the synthesis of space-based and in situ observations. Thus, there should be an expansion of that coverage in the meridional (north-south) direction. This would help to address the tracking of where the heat goes at the end of an El Niño event. A fundamental limitation on monitoring ENSO has also been the absence of sea surface salinity fluxes.

Observations are needed to supplement as well as to verify atmospheric model runs. Retrospective assessments using various data sets (derived from space-based platforms and in situ measurements on land, sea and in the atmosphere) would be used to "fill in" the impacts maps produced in the mid-1980s by Ropelewski and Halpert (1987). For example, looking back at the 1982–83 El Niño and the La Niña events since then suggests that space-based estimates of vegetation provide a useful complement to estimates of rainfall over affected regions like Northeast Brazil. The main conclusion to draw on this point is that the community now possesses new and relatively untapped data sets and sources that can be of use to decision makers for updating our understanding of ENSO impacts on a global scale. Finally, the importance of the location of the 27–28°C isotherm with regard to the onset of active atmospheric convection, suggests that Niño3 may not be the best place to monitor for the onset of La Niña because strong La Niña events tend to develop much further to the western part of the Pacific basin.

In April and May 2000, the Niño3 index switched to positive for the first time since La Niña began. While cooler sea surface temperatures still existed west of the Niño3 region, a number of forecast models called for the La Niña event to be over by summer 2000.

Key points

1. The appearance of depressed sea level height in the western tropical Pacific observed by TOPEX/Poscidon was indicative of a shoaling (less deep) thermocline and signaled the end of 1997–98 El Niño and the possible onset of a cold event (of uncertain magnitude). The majority of this signal took place outside the TOGA-TAO array. Within the context of this event (but consistent with the wind stress curl changes of other events), this suggests that in order to improve forecasts related to La Niña events, (a) a better synthesis of remotely sensed and in situ observations is needed, and (b) the TAO array should be expanded in its coverage in a poleward direction in the western Pacific.
2. Given the "unexpected" behavior (e.g., warming) of the Pacific, Atlantic, and Indian Oceans, simultaneous monitoring of the Indian and Atlantic Oceans must be improved as they influence the effects around the globe of warm and cold extreme ENSO events.
3. The monitoring of changes in ocean color from space (e.g., phytoplankton blooms) west of the Galapagos Islands a couple of months before other key indicators identified changes in oceanic upwelling processes. The identification of such blooms provides hope for an earlier detection than is currently available of changes in an El Niño's strength and, therefore, in the progression of the ENSO cycle.
4. There is not, as yet, much understanding about the role of subtropical SSTs in relation to La Niña events. Expanding the monitoring system to higher latitudes (poleward) in the Pacific would improve this situation.

REFERENCES

Murtugudde, R., S. Signorini, J.R. Christian, A.J. Busalacchi, and C.R. McClain, 1999a: Ocean color variability of the tropical Indo-pacific basin observed by SeaWIFS during 1997–1998. *Journal of Geophysical Research*, 104, 18,351–5.

Murtugudde, R., J.P. McCreary Jr., and A.J. Busalacchi, 1999b: Oceanic processes associated with anomalous events in the Indian Ocean with relevance to 1997–1998. *Journal of Geophysical Research*, 105, 3295–306.

Picaut, J., E.C. Hackert, A.J. Busalacchi, and R. Murtugudde, 2001: Mechanisms of the 1997–98 El Niño-La Niña, as inferred from space-based observations. *Journal of Geophysical Research*, submitted.

Ropelewski, C.F. and M.S. Halpert, 1987: Global and regional scale precipitation patterns associated with the El Niño/Southern Oscillation. *Monthly Weather Review*, 115, 1606–26.

Part 3
Country case studies

Introduction

Michael H. Glantz

In this part of the book several case studies are presented related to La Niña. The cases were drawn from various parts of the globe and from different sectors of society. It is important to keep in mind that La Niña events have seemingly captured less attention from researchers and the public (and certainly the media) than those of its warm counterpart, El Niño. In addition, fewer cold events have occurred since the early 1970s than El Niño events, so that La Niña's impacts on societies and the environment have been less well observed by researchers than those of El Niño. In this regard, it is not so clear how a La Niña's possible impacts would play out in reality. La Niña events come in different intensities, like El Niño's, and so its impacts will likely vary from one event to the next. Some countries and some sectors in society are more vulnerable than others. Yet, some of the La Niña/societal impacts relationships (i.e., teleconnections) can be considered reliable enough for use in critical decision-making situations at various levels of social organization, from individuals to national leaders.

La Niña had been forecast in the early months of 1998 to begin later in the year. A La Niña, in fact, did develop, as was forecast by all forecasting groups. When the El Niño began to decay rapidly (more rapidly than any other at about a degree Celsius a week) in May, forecasters speculated that the expected La Niña had the potential to rapidly become a very intense cold event. The sea surface temperatures in the central Pacific, however, dropped into the normal range for a while, and then it

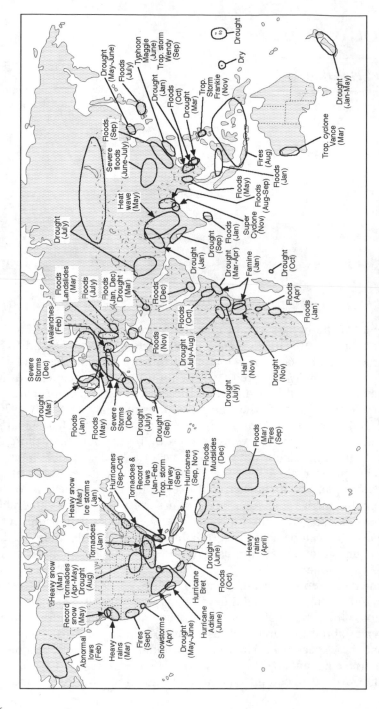

Fig. 3-1 Anomalies in 1999: a La Niña year (Glantz, 2001).

was forecast to continue only as a moderate event through the Northern Hemisphere winter of 1998 and into the spring of 1999. It did.

In late November 1998, a NOAA scientist (personal communication, 23 November) noted that "La Niña is still cooking. The past week has seen some eastward extension of the cold sea surface temperatures into that area that had been sitting just above normal in terms of sea surface temperatures for so many months. It will be interesting to see if the cooling amplifies in the next month or two." It did.

On that same day (23 November 1998), a news release by the BBC News Online Science Editor noted that "somehow El Niño and La Niña are coexisting in the Pacific" and that "the forecasted intensification of La Niña at the end of 1998 has not yet lived up to its billing" because "the size and heat content of this cold pool of water has remained remarkably stable for the past five months" (news.bbc.co.uk:80).

The most recent La Niña event proved to be longer than anticipated, as it continued into the year 2000. The entire year of 1999 witnessed a cold event. Figure 3-1 shows a large number of the notable climate anomalies and extreme events that occurred in 1999. It is important to remember that not every anomaly that occurs in a year in which there is a La Niña in progress is caused by or associated with that event.

As a rule of thumb, one could argue that a review of the numerous impacts associated with the various post-1950 El Niño events could provide some glimpse of what might be *less likely* to occur during a La Niña event. For example, Indonesia is known to be adversely affected by droughts, rainforest fires, and haze during El Niño episodes. This is what the Indonesian government officials fear. When El Niño ends, so too does the likelihood of severe, widespread drought and forest fires (this is because the monsoonal rains return to the region). Thus, as a first approximation of La Niña's impacts in Indonesia, one could surmise that El Niño's feared impacts would not be something to worry about. The same type of review could be applied to climate conditions in other regions that are known to be reliably associated with El Niño (southern Africa, Philippines, Peru, Australia), other things being equal.

REFERENCE

Glantz, M.H., 2001: *Currents of Change: Impacts of El Niño and La Niña on Climate and Society*. Cambridge, UK: Cambridge University Press.

North America

Energy, economics, and ENSO in the United States

*Allan D. Brunner**

Several local climatic anomalies occurred in the United States that have been attributed to the 1997–98 ENSO warm event. This event has been "credited," for example, with the complete absence of hurricanes during the summer and fall of 1997 and with heavy rains and flooding in California as well as in several coastal states bordering the Gulf of Mexico during the Northern Hemisphere winter of 1997–98. Unusually mild weather in the northern and eastern states and unseasonably harsh weather in the southern states have also been associated with the recent El Niño. In early 1998, the media began to note the possibility of an upcoming "cold" event, which reportedly would have the opposite effects to a warm event.

Such adverse fluctuations in local climate are likely to have significant economic consequences for some regions and for some industries in the United States. However, their association with (i.e., teleconnection to) a global and persistent climate anomaly raises an important question for macro-economists: can climate anomalies really have large aggregate effects on the US economy?

An extensive economic literature has been devoted to measuring the effects of changes in local weather (primarily precipitation and temperature) on economic activity (for example, on agricultural production, construction activity, and energy demand). The conventional wisdom seems to be that these effects vary greatly across geographic regions and across industrial sectors but that they do not have important aggregate effects on the US economy. But, if the ENSO warm event/cold event cycle does play a significant role in determining US weather patterns, then it is also possible that ENSO has statistically significant and economically important effects on the US economy.

What economists do know about ENSO

My recent research agenda has had the objective of informing economists and policy makers about the economic consequences of ENSO warm and cold events. The main contribution of this work to the economic literature has been the development of a statistical model that relates ENSO conditions in the tropical Pacific Ocean to state-level average monthly temperature and precipitation readings in the United States. An important feature of this model is the use of continuous measures of ENSO's intensity, e.g., Pacific sea surface temperature and sea level atmospheric pressure anomalies. This is an important improvement, when compared to previous studies that used dummy variables to designate years in which there was unusual climatic activity. Since relatively weak ENSO events were averaged with more severe episodes, the estimated effects were likely biased toward zero and, therefore, toward insignificance.

The dynamics of the model are quite general, allowing for varying effects across states and across time. Using the statistical model, one can answer such questions as "If sea surface temperatures in the Pacific rise unexpectedly by one standard deviation in June, what will be the effect on average monthly temperature and precipitation in August in Alabama?" These effects can then be linked to the previously mentioned economic models that relate changes in temperature and precipitation to changes in US energy consumption or in agricultural production.

Preliminary research results for the effects of ENSO events on energy consumption can be summarized as follows. First, ENSO events have an important statistically identifiable influence on the local climates of the 48 contiguous states. Unexpected changes in ENSO intensity levels affect US temperatures within a few months. ENSO shocks have varying degrees of impact, depending on the month in which they occur, and they have varying effects for different regions across the United States. In addition, ENSO shocks account for a sizeable portion of the monthly variance in US temperatures, although their influence varies from month to month and from state to state. Second, ENSO shocks also have important aggregate effects on US temperatures. For example, during the 1982–83 ENSO event, large regions of the United States were significantly warmer than usual, while other large regions were much cooler. Finally, ENSO shocks also have a substantial aggregate influence on US energy consumption. As a result of the 1982–83 ENSO event, for example, there were $800 million swings in quarterly energy consumption.

What economists don't know about ENSO

There are still many aspects of ENSO events that economists do not fully understand. There is much to be learned before they can make definitive and credible statements and provide meaningful policy advice about ENSO's economic consequences. Many of these questions must be addressed through conversations with physical scientists. Such questions can be grouped into two broad categories – (1) modeling issues, and (2) calibration issues.

1. Very few econometric (statistical) models contain measures or indicators of ENSO activity. How should ENSO events and their associated effects on the global climate be incorporated into these models? I have advocated using continuous measures of ENSO intensity (Brunner, 1998). Is this appropriate? Which measures are the most reliable? Which measures provide the most predictive power for future US or world climatic activity? Is one measure adequate or should multiple measures be used? Does ENSO have nonlinear effects on the world climate? Does a two-degree change in sea surface temperatures have twice the effect as a one-degree change? Does a one-degree increase have exactly the opposite effect as a one-degree decrease?

2. Economists also need a better understanding of ENSO's effects on global climate in order to better calibrate and elucidate their estimated statistical models. Is there a universally accepted taxonomy for identifying ENSO events? How much do we know about the origins of ENSO events? What about the progression and predictability of ENSO events and their influences on climates across the globe, once the ENSO events are identified? Does global warming have any identifiable reliable effect on ENSO events?

Answers to these questions will certainly help economists and other social scientists to build better econometric models and to provide better judgments and policy advice concerning the economic consequences of ENSO events.

Note

*This research was initiated while the author was at the Federal Reserve Board.

REFERENCE

Brunner, A.D., 1998: El Niño and world commodity prices: warm water or hot air? International Finance Discussion Paper (IFDP) no. 608. Recent IFDPs are available on the web at: www.bog.frb.ged.us

La Niña from a Canadian agriculture perspective

Ray Garnett

The pervasive impact of El Niño/Southern Oscillation (ENSO) on world weather patterns in general and on North American weather and climate has been of intense study in the last 15 years or so (Ropelewski and Halpert, 1987, 1989; Kiladis and Diaz, 1989). These and many other studies reported in recent literature have helped identify temperature and precipitation anomalies associated with the El Niño (warm phase of ENSO) in various parts of the world in general and over conterminous United States in particular. The impact of El Niño and its counterpart La Niña (cold phase of ENSO) on Canadian temperature and precipitation patterns has been documented in two recent studies (Shabbar et al., 1997). In general it can be stated that El Niño brings warmer and drier weather and La Niña colder and wetter weather over western Canada, specifically during the winter following the onset of an ENSO extreme event.

The impacts of El Niño and La Niña on Canadian agriculture and, in particular, on Canadian wheat yields, have been studied by Garnett and Khandekar (1992), among others. In general, it has been found that El Niño years tend to favor the wheat yield while La Niña years tend to be unfavorable for the wheat yield. Specfically, when the sea surface temperatures in the central and eastern equatorial Pacific are above normal during winter and early spring, the rainfall over the Canadian prairies during the months of June and July is above normal which in turn favors the wheat yield (Garnett and Khandekar, 1995; Le Faivre et al., 1997;

Hsieh et al., 1999). One of the conclusions reached by Hsieh and colleagues in 1999 was that La Niña conditions in the equatorial region have a much more significant influence on Canadian wheat yield than El Niño conditions. Their study suggested that low yields over the Canadian prairies may be considerably more predictable than high yields from the sea surface temperature anomaly (SSTA) data, thanks to a better signal to noise ratio.

The SST distribution in the central and eastern equatorial Pacific influences the central and north Pacific atmospheric flow patterns as documented in a landmark paper by Wallace and Gutzler (1981) who identified the PNA (Pacific North American) flow patterns for which suitable PNA indices have been developed. In a recent study (Garnett et al., 1998) the utility of ENSO and PNA indices for long-lead forecasting of summer weather over the Canadian prairies has been demonstrated. A simple technique involving accumulated values of the Pacific North American Teleconnection index has consistently provided general guidance in foreshadowing June and July temperature since 1994.

Dr. Khandekar and the writer of this chapter were able to forecast the Canadian prairie winters of 1995–96 (La Niña), 1996–97 (La Niña) and 1997–98 (El Niño). Incorrect forecasts were made for the Canadian prairie winters of 1998–99 and 1999–2000 based solely on La Niña conditions. Future winter forecasts will need to incorporate factors such as the North Pacific and North Atlantic oscillations in a multivariate approach, as suggested by Garnett and Khandekar (1992).

El Niño, La Niña, and Canadian wheat yields

As mentioned earlier, El Niño years are found to be favorable and La Niña years unfavorable for wheat yield over the Canadian prairies. After observing the global impact of the strong 1982–83 El Niño/Southern Oscillation as an analyst in the Canadian Wheat Board's Weather and Crop Surveillance Department, the writer hypothesized that El Niño is favorable for the spring wheat crop over the Canadian prairies and unfavorable for the Australian wheat crop, while La Niña is unfavorable for the Canadian spring wheat crop and favorable for the Australian wheat crop. When the yield series of the two countries are overlaid, one can detect the El Niño/Southern Oscillation events in the two yield series. Like tree rings, wheat yields are a proxy variable for climate. In essence, these two countries are climatic twins with an inverse statistically significant correlation with an r value of $-.4$ for the period 1960–98. These two crop regions exist on opposite sides of the El Niño/Southern Oscillation phenomenon.

Impact of other large-scale features on Canadian grain yield

Garnett and Khandekar (1992) also analyzed the impact of large-scale features like the equatorial stratospheric quasi-biennial wind oscillation (or the QBO, as it is commonly known), Eurasian winter snow cover and Indian summer (June–September) monsoon rainfall. Interestingly, the westerly (easterly) phase of the QBO is found to influence the Indian summer monsoon rainfall favorably, while the Indian summer monsoon rainfall is found to be strongly and inversely correlated with US corn yield, but only weakly (and inversely) correlated with Canadian spring wheat yield.

The influence of ENSO, QBO, and Eurasian winter snow cover on the Indian monsoon has been studied by Khandekar (1996) and Khandekar and Neralla (1984), among others. This and many other studies have shown that in general El Niño (La Niña) tends to be unfavorable (favorable) for the Indian summer monsoon. By assessing the general strength of the Indian monsoon during the mid-summer season in a given year, it may be possible to foreshadow the US corn yield and the Canadian spring wheat yield with a lead time of one to three months. Serious droughts like 1983 and 1988 in the US corn belt have occurred simultaneously with a flood monsoon in India. The severe drought of 1961 on the Canadian prairies also occurred during a severe flood monsoon in India. During the 1930s when dust bowl conditions prevailed over the Canadian prairies there were three flood monsoons and seven normal monsoons (Khandekar and Neralla, 1984).

A simple technique of accumulating Niño3 sea surface temperature anomalies since September provides very general guidance as to June and July precipitation over the Canadian prairies (Garnett et al., 1998).

Concluding remarks

El Niño and La Niña appear to provide a strong climatic signal on the climate and agriculture of western Canada. Recent studies have documented this impact through empirical and correlation analysis. Continuation of such studies will help improve our capability for long-lead forecasting of grain yields over the Canadian prairies with a lead time of three to six months.

REFERENCES

Garnett, E.R. and M.L. Khandekar, 1992: The impact of large-scale atmospheric circulations and anomalies on Indian monsoon droughts and floods and on world grain yields – statistical analysis. *Journal of Agricultural and Forest Meteorology*, 61, 113–28.

Garnett, E.R. and M.L. Khandekar, 1995: *Proceedings of the Long-Range Weather and Crop Forecasting Working Group Meeting II*, 21–23 March 1995, Winnipeg, Manitoba.

Garnett, E.R., M.L. Khandekar, and J.C. Babb, 1998: On the utility of ENSO and PNA indices for the long-lead forecasting of summer weather over the crop growing region of the Canadian prairies. *Journal of Theoretical and Applied Climatology*, 60.

Hsieh, W.W., B. Tang, and E.R. Garnett, 1999: Teleconnections between Pacific sea surface temperatures and Canadian prairie wheat yield. *Agricultural and Forest Meteorology*, 96, 209–17.

Khandekar, M.L., 1996. El Niño/Southern Oscillation, Indian monsoon and world grain yields – A synthesis. In M.I. El Sabh et al. (eds.), *Land-based and Marine Hazards* (Advances in Natural and Technical Hazards Research), 79–95. Netherlands: Kluwer Academic.

Khandekar, M.L. and V.R. Neralla, 1984: On the relationship between the sea surface temperatures in the equatorial Pacific and the Indian monsoon rainfall. *Geophysical Research Letters*, 11, 1137–40.

Kiladis, G.N. and H.F. Diaz, 1989: Global climate anomalies associated with extremes in the Southern Oscillation. *Journal of Climate*, 2, 1069–90.

Le Faivre, L., M.L. Khandekar, and E.R. Garnett, 1997: *Proceedings of the Long-Range Weather and Crop Forecasting Working Group III*, 20–21 October 1997, Dorval, Quebec, Canada.

Ropelewski, C.P. and M.S. Halpert, 1987: Global and regional scale precipitation patterns associated with El Niño/Southern Oscillation. *Monthly Weather Review*, 115, 1606–26.

Ropelewski, C.P. and M.S. Halpert, 1989: Precipitation patterns associated with the high index phase of the Southern Oscillation. *Journal of Climate*, 2, 268–84.

Shabbar, A., B. Bonsal, and M.L. Khandekar, 1997: Canadian precipitation patterns associated with the Southern Oscillation. *Journal of Climate*, 10, 3016–27.

Wallace, J.M. and D.S. Gutzler, 1981: Teleconnections in the geopotential height field during the Northern Hemisphere winter. *Monthly Weather Review*, 109, 784–812.

The impact of ENSO on Canadian climate

Amir Shabbar

The effects in North America of El Niño/Southern Oscillation (ENSO) events are usually strongest during the Northern Hemisphere winter. Recent investigations into the impact of ENSO on Canadian climate also demonstrate that the tropical influence is strongest during the cold season. In this study, Canadian temperature and precipitation and Northern Hemisphere mid-tropospheric data have been statistically analyzed in the context of El Niño, La Niña, and non-ENSO years (Shabbar and Khandekar, 1996; Shabbar et al., 1997). The analysis incorporated 25 moderate to strong El Niño events and 17 moderate to strong La Niña events that have occurred in the twentieth century. Differences in the Pacific North American (PNA) oscillation and, in particular, the height anomaly over western Canada, provide the strongest ENSO signal over North America. It also explains the significant temperature and precipitation responses over Canada. However, results indicate that the effects of La Niña are not exactly opposite to those associated with El Niño. Given that the nature of ENSO events introduces a lead time in the climate over Canada, the established relationships in this study have been used, through the canonical correlation analysis (CCA) technique, to produce long-lead forecasts of Canadian temperature and precipitation. Recent ENSO events have shown that these forecasts provide usable skill.

Recent ENSO impacts over Canada

Each El Niño and La Niña has its own distinct characteristics. Examples of a few recent major impacts are listed below.

1. *El Niño*: With the exception of the northeastern Arctic, winter temperatures were 2 to 8 °C above normal, and precipitation was generally one-half of the normal amount in association with the 1997–98 El Niño. The warm, dry weather contributed to severe brush fires in Alberta in December 1997. A temporary northeastward displacement of the unusually strong El Niño-related sub-tropical jet stream is also believed to have played a key role in the disastrous ice storm over eastern Ontario and southwestern Quebec in January 1998. During the 1982–83 and 1997–98 warm events, the mild winter weather significantly reduced snow removal budgets for municipalities. Heating costs were 5 to 15 percent below normal for homeowners and businesses (an average savings of C$200 per household).

2. *La Niña*: During the 1988–89 event, a build up of bitterly cold air in the high Arctic was responsible for the highest pressure reading ever recorded in North America: 107.5 kilopascals at Northway, Alaska. All-time high-pressure readings were also set in many stations on the Canadian prairies. During early February 1989, Vancouver experienced its longest cold spell, as the overnight temperatures dropped below −10 °C. Frigid air pushed as far south as southern Texas and even blanketed southern California with snow. During the 1995–96 La Niña, most of Canada suffered from a particularly long winter. Even in southern Canada, most locations had continuous snow cover from October to March. Winter temperatures were as much as 11 °C below normal in Winnipeg (Manitoba); the second coldest on record. In addition, overnight temperatures remained below a frigid −30 °C for 19 days in a row! Higher than normal snowfall in the Red River Valley contributed to an extensive spring flooding in southern Manitoba and in North and South Dakota in the United States.

Temperature impact

Our analysis shows that statistically significant positive surface temperature anomalies spread eastward from the west coast of Canada to the Labrador coast during the late fall to early spring (November through May), following the onset of El Niño episodes. The accompanying temperatures in the lower troposphere show a transition from the PNA pattern to the tropical/Northern Hemisphere (TNH) pattern over the North American sector during the same period. Conversely, significant negative

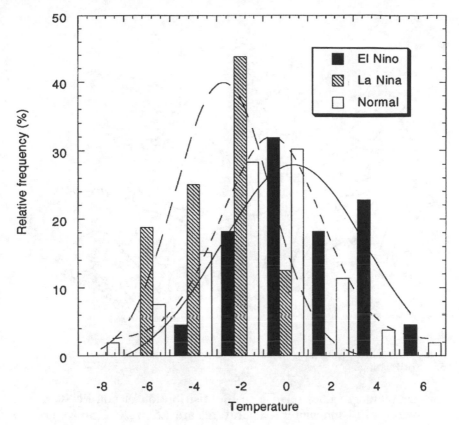

Fig. 3-2 Temperature distribution over western Canada – typical winter precipitation response following the onset of La Niña. Values shown are standardized anomaly.

surface temperature anomalies spread southeastward from the Yukon and extended into the upper Great Lakes region, during the winter season following the onset of La Niña episodes. The lower tropospheric temperatures show a negatively phased PNA-like pattern in early winter, which weakens considerably by May of the following year. Thus, while western Canadian surface temperatures are influenced during both phases of ENSO, those over eastern Canada are affected only during El Niño. The impact of ENSO on the Canadian surface temperatures is the strongest during the winter season and nearly disappears by the spring (April and May). In both El Niño and La Niña years, the largest temperature anomalies are centered over two separate regions: one over the Yukon and the other just west of Hudson Bay. Over western Canada, the mean temperature distribution of the El Niño (La Niña) years is shifted toward

Fig. 3-3 Western Canada – typical winter precipitation response following the onset of La Niña.

warmer (colder) values relative to the distribution of non-ENSO years. On average, El Niño winter temperatures are 1.5 to 2.0 °C above normal over western Canada, while La Niña winter readings are 2.0 to 3.0 °C below normal over the same locations.

In keeping with the historical pattern, the 1997–98 El Niño brought above-normal temperatures. For the country as a whole, it was the second-warmest winter in the instrumental record. Even in the face of the La Niña during the winters of 1998–99 and 1999–2000, the temperatures over Canada continued to show above-normal values. This behavior is consistent with the strong warming trend observed over the land areas of the Northern Hemisphere in recent years.

Precipitation impact

A detailed investigation of the spatial and temporal behavior in precipitation responses over Canada shows significant responses over southern Canada, during the first winter following the onset of the ENSO events. Composite and correlation analyses indicate that precipitation over a large region of southern Canada extending from British Columbia, through the Prairies, and into the Great Lakes region is significantly influenced by the ENSO phenomenon. The results show a distinct pattern of negative (positive) precipitation anomalies in this region during the first winter following the onset of El Niño (La Niña) events. Statistical and field significance of the responses are established by non-parametric and Monte Carlo procedures. The significant precipitation anomalies can be explained by the associated mid-tropospheric flow pattern, which, following the onset of El Niño (La Niña) events, resembles the positive (negative) phase of the PNA pattern. While the temperature anomalies for both El Niño and La Niña show a west-to-east progression starting in the autumn of the onset year and persisting through the following spring, no progression of precipitation anomalies is evident in our analysis. Furthermore, the areas encompassed by significant temperature anomalies differ from those of precipitation. Over the southern Canadian region, an investigation of the intra-seasonal variability shows no apparent difference in the intensity of precipitation. In addition to the ENSO signal, winter precipitation over Canada also appears to be influenced by decadal-scale oscillations in the Pacific Ocean. There is considerable interannual variability in both temperature and precipitation responses associated with individual ENSO events. In opposition to the historical La Niña precipitation patterns, below-normal amounts of precipitation were observed over the Prairie provinces during the La Niña winters of 1998–99 and 1999–2000.

REFERENCES

La Niña from a Canadian Perspective – www.msc-smc.ec.gc.ca/laNiña/ (as of November 2001).

Shabbar, A. and M. Khandekar, 1996: The impact of El Niño-Southern Oscillation on the temperature field over Canada. *Atmosphere-Ocean*, 34(2), 401–16.

Shabbar, A., B. Bonsal, and M. Khandekar, 1997: Canadian precipitation patterns associated with the Southern Oscillation. *Journal of Climate*, 10, 3016–27.

The effects of ENSO on California precipitation and water supply

Maurice Roos

Background

The primary activity of the California Department of Water is water supply forecasting, including snowmelt runoff and, jointly with the NWS California-Nevada River Forecast Center, flood forecasting on the major rivers of northern California. The former is a longer-range forecast of about six months into the future (or to the end of the water year, which is on 30 September), which could employ longer-range weather forecasts if sufficient skill was demonstrated. The precipitation projection used for flood forecasting is much shorter (in hours up to a couple of days) and probably not a good target for eastern tropical Pacific warm or cold event forecasts. However, there may be some definitive statements on the change in probabilities of flood events during a season which can be developed from El Niño and La Niña conditions which could be useful in emergency planning.

Water supply forecasts are used by water project operators to schedule reservoir operations and water system deliveries. In general, reservoir management is very conservative so as to avoid shortage, that is, a need for reductions in planned deliveries after the delivery season has started. Changes in delivery to State Water Project water contractors will occur only if the 99 percent probable amount surpasses the 90 percent probability value projected in December. For example, if the conventional 1 December, 90 percent probability amount was 11 million acre feet of

Sacramento River system runoff, the predicted runoff as of 1 January (or any subsequent date) must exceed this amount at the 99 percent probability to change the allocation.

Goals

What information is available on relationships between La Niña and worldwide weather patterns, especially those which might affect California during the winter wet season? There are indications from past such events of dry winters in southern California and Arizona, but a much more mixed outlook for northern California. In fact, some of our worst floods have come in La Niña years. We wonder whether the major river flood threat is sufficiently high to warrant special preparations over and above those made for every winter. Would such a forecast be convincing enough to ask the legislators and Governor of the State of California for some pre-deployment funds for flood fighting materials and activities as was done in 1997, based on the El Niño-related forecasts?

From a water-operation study perspective, there are two thresholds of long-range forecasting which would be useful. The first is a reliable forecast of a wet or dry season early in the water year. (The water year is the 12-month period that begins 1 October and extends through 30 September.) Generally, in this region not much precipitation occurs after mid-April, so the seasonal forecast in November or on 1 December only needs to extend by about five months. Such a forecast would still be quite valuable on 1 January, which is at about the 35 percent mark in the accumulation season. By early February many of the major operational planting decisions in agriculture are made, with mostly small adjustments being made as a result of later weather events. At that point, the major benefit would shift to a reliable forecast of the next wet season (a 15-month forecast), which would affect the amount of reservoir carryover to save for a possible dry winter during the next season. This is especially true if the current year is dry and users need to weigh the options of using most of their stored water in the current season or taking some losses to ensure better water supply during the following year.

La Niña impacts on the Pacific Northwest

Nathan Mantua

Recent advances in climate science have demonstrated the possibility for improved skill in forecasting tropical El Niño and La Niña conditions as much as one year in advance (Latif et al., 1998). However, these predictions are not so good that we can expect to forecast well in advance ENSO extreme event intensities with skill. This chapter explores the potential for extending La Niña predictions to skillful climate forecasts for the North American Pacific Northwest region (Figure 3-4 (c)). The potential societal benefits from improved climate forecasts are also addressed. The key question is this: should we expect that Pacific Northwest public and private agencies, as well as individuals, will be able to use La Niña-related climate forecasts to mitigate expected losses and/or exploit expected opportunities?

The brief data analysis presented in this chapter documents empirical relationships between the tropical La Niña and some features of Pacific Northwest (PNW hereafter) climate. Several recent studies have found statistically significant correlations between La Niña events and PNW climate (e.g., Redmond and Koch, 1991; Pulwarty and Redmond, 1997; Hamlet and Lettenmeier, 1999; Cayan et al., 1999; Miles et al., 2000), although the results of some earlier studies did not find consistent relationships between PNW climate and La Niña (Ropelewski and Halpert, 1986). The analysis done in this study differs from that of the earlier studies in several respects. For example, Ropelewski and Halpert (1986) focused on all La Niña episodes above a specified threshold, while here the historical record is segregated into subjectively defined "strong" and "moderate" intensity La Niña events; Redmond and Koch (1991) focused on linear correlations between the Southern Oscillation Index (SOI) and western US temperature and precipitation, while here the focus is on estimated probability density functions and composite monthly mean anomalies. The results from the analysis that follows highlight some of the limitations of linear statistical analysis (e.g., linear correlation studies), and how important information may be lost by ignoring the possibility for non-linear empirical relationships.

Given the present-day capabilities in monitoring and predicting La Niña (and El Niño) variability at seasonal to interannual time scales, there appears to be great potential for predicting aspects of PNW climate during periods with strong La Niña conditions. With anomalous climate conditions come potential impacts on climate-sensitive sectors and activities, a situation which allows for climate forecasts to be extended to value-added "resource-impacts forecasts."

The data and analysis used in this study are briefly described. PNW climate anomalies associated with twentieth-century La Niña events are then described. La Niña-related impacts in the PNW are discussed, followed by a general discussion of the use and utility of climate information.

Data and analysis

In the analysis that follows, relationships between La Niña and PNW climate rely upon subjectively defined criteria for identifying "strong" and "moderate" intensity La Niña episodes in the past century. Here, the October–March averaged Niño3.4 index of equatorial Pacific sea surface temperatures is used as the defining index for El Niño/La Niña events. Niño3.4 data were obtained from NOAA's Climate Diagnostics Center for the 1950–2000 period, while Kaplan et al.'s (1998) Optimally Smoothed Index was used for the period 1895–1949. Anomalies are computed with respect to the 1950–95 climatology. October–March means are chosen to limit our focus on the calendar months that historically have shown the strongest teleconnections between tropical Pacific and North American climate (e.g., Horel and Wallace, 1981; Ropelewski and Halpert, 1986; Trenberth et al., 1998). For a detailed discussion of the relative merits of the Niño3.4 index, see Trenberth (1997).

"Strong" La Niña episodes are defined as those having a normalized October–March Niño3.4 value more negative than −1.2; "moderate" intensity La Niña episodes are defined as those years having October–March Niño3.4 values between −0.5 and −1.2; "non-ENSO" years are identified as those years with October–March averaged Niño3.4 values

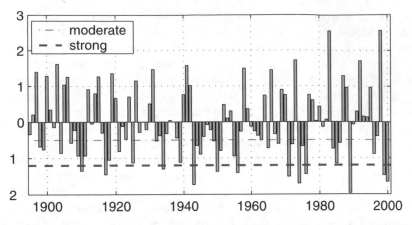

Fig. 3-5 Normalized October–March averaged Niño3.4 time series for 1895–2000. Each year indicates averaged Nino3.4 values for the period October (year −1) through March (year 0). Horizontal dashed lines are drawn at −0.5 and −1.2 to indicate thresholds used to identify "moderate" and "strong" intensity La Niña events.

between +/−0.5 (Figure 3-5). Based on these criteria, there are 10 strong and 21 moderate intensity La Niña events, and 32 non-ENSO years, in the 1895–1997 period of record (see Table 3-1). Note that using other (equally subjective) criteria to identify La Niña events and non-ENSO years yields different results (cf. Trenberth, 1997; Rasmusson and Carpenter, 1982; Wolter and Timlin, 1998).

Climate data used in this study focused on the hydroclimate variability in the PNW region. Here the PNW region is defined as the combined three-state region of Idaho, Oregon, and Washington, plus the remaining parts of the Columbia River Basin (Figure 3-4). Data examined include: surface air temperatures and precipitation from Oregon, Idaho, and Washington US Climate Division Data (Karl et al. 1986); "naturalized" streamflow from the Columbia River at The Dalles, Oregon, which integrates the hydrology for much of the Columbia River Basin (outlined in Figure 3-4) and serves as an informative proxy for PNW hydrology as a whole; and snow depth from Paradise Ranger Station on Mount Rainier, Washington. Anomalies for each data set are computed with respect to the 1950–95 climatology. More complete information on these data is listed in Table 3-2.

In our first examination of the relationships between La Niña events (moderate and strong) and PNW climate, estimated probability density functions (PDFs) are computed for each sub-sample of the historic record. Using the second analysis method, composites of monthly mean and seasonal mean data are constructed for selected PNW climate measures.

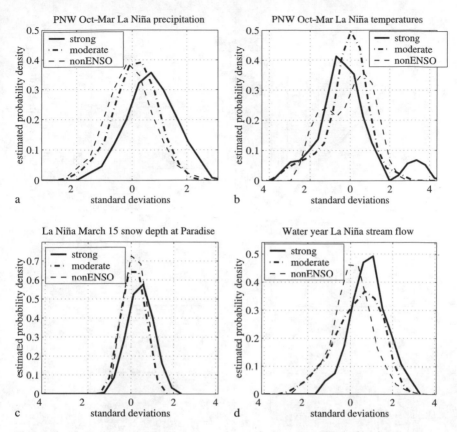

Fig. 3-6a–d Estimated October–March surface probability density functions (PDFs), regionally averaged for the PNW (Idaho-Oregon-Washington). The strong La Niña sample PDF is given with the heavy solid curve; the moderate La Niña sample PDF is given with the dash-dot curve; the non-ENSO PDF is given with the light dashed curve (see key).

receive less than the PNW regional average precipitation as a whole, the composite rainfall anomalies for strong La Niña years are small yet still positive (<5 cm over the six-month period).

The temporal characteristics of La Niña-related PNW climate anomalies are shown with composites of monthly mean data for surface temperatures, precipitation, snow depth at Paradise Ranger Station, and Columbia River stream flows in the four panels of Figure 3-8. Precipitation tends to be especially above average during November–January, while the composite temperature anomalies are below average in October, February, March, and June, but above average in January and April. This tendency for cool-to-average temperature and above average

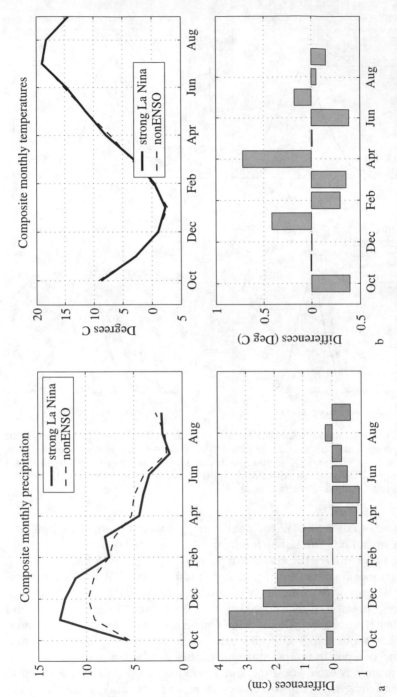

Fig. 3-8a–d Composite monthly mean climate anomalies for strong La Niña year and non-ENSO subsamples; bar graphs indicate the differences between the composites (La Niña/non-ENSO). Line patterns follow the convention used in Figure 3-6.

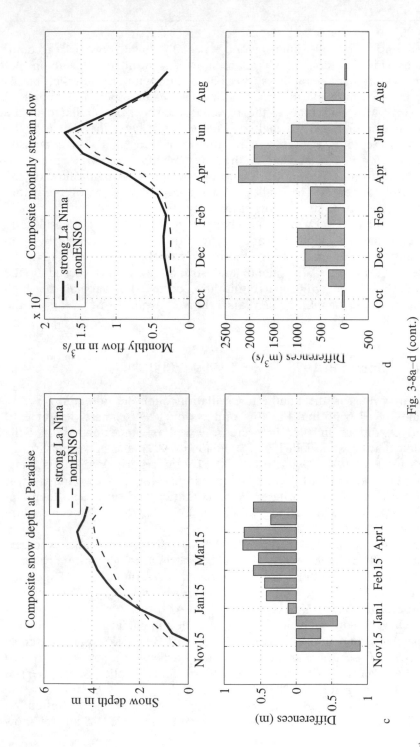

Fig. 3-8a–d (cont.)

precipitation in fall-winter favors especially high snowpack accumulations (Figure 3-8c); the abundant snowpack ultimately results in elevated water-year stream flows throughout the region. In snowmelt basins like the Columbia Basin, stream flows tend to be especially anomalous in April, May and June as the regional snowpack accumulated in the fall and winter months is melted (Figure 3-8d). Over the course of the October–September "water year," the strong La Niña year composite precipitation is 109 percent of the "non-ENSO" composite precipitation; the strong La Niña year composite water-year stream flow on the Columbia River at The Dalles, Oregon, is 114 percent of the "non-ENSO" composite stream flow at this location.

Cayan et al. (1999) find that changes in the PDFs for monthly and seasonal mean precipitation and stream flow are accompanied by similar shifts in daily extremes for these two variables, respectively. In the Pacific Northwest, daily rainfall and streamflow events exceeding the ninetieth and ninety-fifth percentile have occurred at anomalously high frequencies, during La Niña years relative to non-ENSO years in the historical record.

PNW climate in 1998–99 and 1999–2000

Recent swings in the Southern Oscillation offer the potential for case studies of El Niño and La Niña impacts on PNW climate. Extreme El Niño conditions in 1997–98 were followed by an extended La Niña period that began in mid-1998 (McPhaden, 1999), and persisted through the winters of 1998–99 and 1999–2000. By the criteria used in this study, October 1998–March 1999 and October 1999–March 2000 both qualify as "strong" events. Composite PNW precipitation and surface air temperature maps for these periods, based on US Climate Division data, are shown in Figure 3-9 (c).

During October 1998–March 1999 PNW climate anomalies were exceptional. The October 1999–March 2000 period brought unusually heavy precipitation to western Washington and western Oregon. Especially large amounts of precipitation, in many cases exceeding 150 percent to 250 percent of seasonal averages, fell on west-facing slopes throughout the PNW. Record precipitation was recorded in climate divisions for the Washington and Oregon coasts. The Mount Baker ski area, located in the North Cascades near the US-Canada border, set a new *world record* for the highest ever-recorded annual snowfall (October–September) in 1998–99 with a total of 28.96 m (1,142 inches) – it is notable that the previous world record was recorded at Paradise Ranger Station on Washington's Mt. Rainier, when this location recorded 28.5 m of snow fall during the 1971–72 "moderate" La Niña period (Bell et al., 2000). While precipitation and snowpack was well above average in 1998–99 for the PNW region, temperatures were generally near average in the western PNW and ~1 °C above the 1950–95 average in the eastern part of the PNW (top panels of Figure 3-9).

PNW climate anomalies during October 1999–March 2000 were characterized by well above-average temperatures and near-average amounts of precipitation (bottom panels of Figure 3-9). Surface temperature anomalies during this period ranged from slightly above-average values near the coast to as much as 1.5 °C above average in the eastern part of the region. Seasonal accumulations of snowpack were similar to the October–March precipitation anomalies, with slightly above-average amounts in Washington's Cascade and Olympic Mountains, and near-average amounts in Idaho's Rocky Mountains.

La Niña-related climate impacts in the Pacific Northwest

Changes in the Pacific Northwest's hydrology produce a variety of impacts on a wide range of natural resource sectors. For instance, the tendency for above-average water-year streamflows leads to a reduced likelihood for conflicts over water allocation between competing stakeholders in the Columbia River Basin. Likewise, an increased probability for an abundant water supply suggests a similar probability shift towards relatively good in-stream habitat for juvenile anadromous fish (including Pacific salmon) during the typically low-flow, late summer and fall months. The reduced frequency of exceptionally low precipitation, low snowpack, and low stream flow years suggests a reduced likelihood for drought and water-supply shortfalls. On the other hand, the increased risk of extremely high October-to-March streamflow events leads to an increased risk for property damage (flooding) and relatively poor survival of incubating salmon eggs (via scouring of gravel nests).

Because of changes in the preferred character of the winter storm track, strong La Niña years have also brought an increased frequency of low-elevation snowfall events in the western lowlands of the Pacific Northwest (Figure 3-10a). Such storms often produce large financial losses in urban areas like Portland (Oregon) and Seattle (Washington). At the same time, the tendency for abundant mountain snowpack and a low elevation freezing level is welcome news for ski enthusiasts and water managers.

Finally, the tendency for anomalously cool October-to-March surface temperatures is also true for coastal ocean temperatures (Figure 3-10b, cf. Harc et al., 1999). Anomalous cooling of the coastal ocean (along with

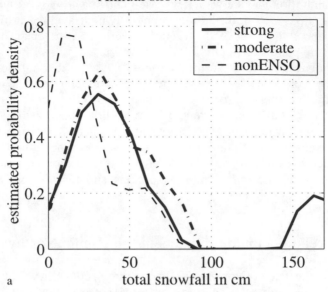

a

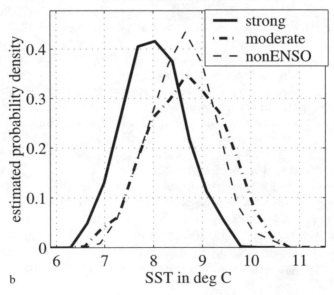

b

Fig. 3-10a, b Estimated PDFs for PNW low elevation snowfall and coastal ocean temperatures.

a suite of related changes to the upper ocean environment) tends to provide favorable conditions for high biological productivity for a number of species, including top-level predators like sea birds, marine mammals, and Pacific salmon.

Discussion

Generally speaking, periods with strong La Niña conditions have been associated with anomalously wet and near to below average temperature winter climate conditions in the PNW. The combination of simultaneous cool or average temperatures and above-average precipitation in the winter season results in an amplified hydrologic response through an enhanced seasonal accumulation of snowpack. The subsequent snowmelt runoff carries the winter climate anomalies into spring and summer stream flows. Conversely, moderate La Niña conditions have not been associated with strong and consistent climate anomalies in the PNW.

Assuming there is a demonstrated capability to predict climate anomalies at seasonal to interannual time scales, can we assume that society as a whole will benefit from this technology? The societal response to the climate anomalies attributed to the 1997–98 El Niño episode suggests that the use of climate information may be as unpredictable and mysterious as the climate system itself.

During the 1997–98 El Niño, the climate forecasting and societal impacts communities were faced with a range of challenges. These included purely scientific questions of climate forecast accuracy, to the flow of conflicting climate information from multiple sources, and then to societal constraints imposed by such things as institutional barriers to the use of climate information, or inequalities in the political or economic power of different users. Generally speaking, the problems in utilizing El Niño-based climate forecasts are identical to those that will limit the use of La Niña-related climate information.

Studies conducted over the past five years by the Climate Impacts Group at the University of Washington find that, in spite of the difficult hurdles listed above, there are relatively fast and inexpensive ways to improve the utility and value of climate forecasts. One route is to extend climate forecasts into *impacts* forecasts. In this case, impacts are viewed as the consequences of an increased probability of physical climate anomalies such as an unusually heavy snowpack. The impacts outlook would take this information and extend it into a prediction for something like an increased probability for an outstanding ski season. The important distinction here is that the climate information is translated into a language that is understandable to a wider audience than that of the climate prediction community.

A second and complimentary route is to develop and maintain an open dialogue with the so-called *user community*. In many cases, the climate research and prediction community generates many products that are largely unintelligible and; therefore, useless to both the general public and decision makers in climate-sensitive sectors. Increased utility of climate information appears to be closely related to value-added information, translation, and directed application to specific issues. Presently, there does not appear to be a lack of climate information, but more likely there are more data and analyses than the non-specialist can digest. It is hoped that concerted efforts to foster collaboration between the climate research, impacts, and user communities will quickly extend the use and usability of climate information.

REFERENCES

Bell, G.D., M.S. Halpert, R.C. Schnell, R.W. Higgins, J. Lawrimore, V.E. Kousky, R. Tinker, W. Thiaw, M. Chelliah, and A. Artusa, 2000: Climate Assessment for 1999. *Bulletin of the American Meteorological Society*, 81, S1–S50.

Cayan, D.R., K.T. Redmond, and L.G. Riddle, 1999: ENSO and hydrologic extremes in the Western United States. *Journal of Climate*, 12, 2881–93.

Hamlet, A.F. and D.P. Lettenmeier, 1999: Columbia River streamflow forecasting based on ENSO and PDO climate signals. *American Society of Civil Engineers*, 125, 333–41.

Hare, S.R., N.J. Mantua, and R.C. Francis, 1999: Inverse production regimes: Alaskan and west coast salmon. *Fisheries*, 24(1), 6–14.

Horel, J.D. and J.M. Wallace, 1981: Planetary scale atmospheric phenomena associated with the Southern Oscillation. *Monthly Weather Review*, 109, 813–29.

Kaplan, A., M. Cane, Y. Kushnir, A. Clement, M. Blumenthal, and B. Rajagopalan, 1998: Analyses of global sea surface temperature 1856–1991. *Journal of Geophysical Research*, 103, 18,567–89.

Karl, T.R., C.N. Williams Jr., P.J. Young, and W.M. Wendland, 1986: A model to estimate the time of observation bias associated with monthly mean maximum, minimum and mean temperatures for the United States. *Journal of Climate and Applied Meteorology*, 25, 145–60.

Latif, M., D. Anderson, T. Barnett, M. Cane, R. Kleeman, A. Leetmaa, J. O'Brien, A. Rosati, and E. Schneider, 1998: A review of the predictability and prediction of ENSO. *Journal of Geophysical Research*, 103, 14,375–94.

McPhaden, M.J., 1999: Genesis and evolution of the 1997–98 El Niño. *Science*, 283, 950–4.

Miles, E.L., A.K. Snover, A.F. Hamlet, B. Callahan, and D. Fluharty, 2000: Pacific Northwest Regional Assessment: The impacts of climate variability and climate change on the water resources of the Columbia River Basin. *Journal of the American Water Resources Association*, 36, 399–420.

Pulwarty, R.S. and K.T. Redmond, 1997: Climate and salmon restoration in the

Columbia River Basin: The role and usability of seasonal forecasts. *Bulletin of the American Meteorological Society*, 78, 381–97.

Rasmusson, E.G. and T.H. Carpenter, 1982: Variations in tropical sea surface temperature and surface wind fields associated with the Southern Oscillation/El Niño. *Monthly Weather Review*, 110, 354–84.

Redmond, K.T. and R.W. Koch, 1991: Surface climate and streamflow variability in the western United States and their relationship to large-scale circulation indices. *Water Resources Research*, 27, 2381–99.

Ropelewski, C.F. and M.S. Halpert, 1986: North American precipitation and temperature patterns associated with the El Niño/Southern Oscillation (ENSO). *Monthly Weather Review*, 114, 2352–62.

Ropelewski, C.F. and M.S. Halpert, 1989: Precipitation patterns associated with the high index phase of the Southern Oscillation. *Journal of Climate*, 2, 268–84.

Silverman, B.W., 1986: *Density Estimation for Statistics and Data Analysis*. Chapman and Hall, London.

Trenberth, K.E., 1997: The definition of El Niño. *Bulletin of the American Meteorological Society*, 78, 2771–7.

Trenberth, K.E., G.W. Branstator, D. Daroly, A. Kumar, N.C. Lau, and C. Ropelewski, 1998: Progress during TOGA in understanding and modeling global teleconnections associated with tropical sea surface temperatures. *Journal of Geophysical Research*, 103, 14,291–324.

Wolter, K. and M.S. Timlin, 1998: Measuring the strength of ENSO – how does 1997/98 rank? *Weather*, 53, 315–24.

ENSO forecasts and fisheries

Warren Wooster

Despite claims to the contrary, the ability to forecast an "ENSO" event (i.e., extreme changes in equatorial conditions) may at present have only limited value in forecasting fishery production, especially within a given region or ecosystem distant from the equator, such as the eastern North Pacific. Fishery production is variable on various time scales, from seasons to several decades and longer, and the causes of such variability include not only environmental variability but also the effects of fishing pressure and the responding internal dynamics of ecosystems.

There is, of course, increasing evidence for the importance of environmental variability, for example, in the similarity in the patterns of environmental and fish abundance variability in one or another region, to such an extent that fishing effects may seem trivial. Well-known cases include large-scale changes in the catches of sardine (Kawasaki, 1983) and North Pacific salmon (Beamish and Bouillon, 1993; Francis and Hare, 1994). Other evidence comes from the analysis of fish-scale deposits in the Santa Barbara Basin (Baumgartner et al., 1992), showing large-scale explosions and collapses of the sardine and anchovy populations during the past 1,700 years, long before the existence of fishing activities in the region.

It has become commonplace to attribute various consequences, for example, changes in distribution or abundance of fish stocks, to El Niño (the same logic would apply to La Niña). A problem with this attribution is that any unusual warming (or cooling) is labeled El Niño (or La Niña) even when far from the equator, whether or not it is linked to the tropical event. Not only the causal linkages are sometimes questionable, but the timing and intensities of equatorial and extra-tropical occurrences may be very different. At higher latitudes, the causative factors of distribution and abundance changes are more probably local than remote. For example, unusual warming in the Gulf of Alaska or eastern Bering Sea is more likely to impact recruitment directly than is the distant tropical event that may (or may not) have been the cause of that warming (Hollowed et al., 1998).

If the relationship were established between El Niños (and La Niñas) and local environmental conditions, favorable or otherwise to a given species, a forecast of the distant cause might lead to a useful forecast of production of that species. It must be recognized, however, that the more important effects on fish production may well be those of decadal or longer time scales rather than those on the interannual, i.e., El Niño or La Niña, time scale.

In general, the mechanisms of the interaction between environment and fish stock variations are poorly understood and are the subject of a variety of hypotheses, mostly based on corrrelations rather than on the working out and identification of mechanisms. Explanations for the lack of understanding include the traditional mindset of fishery scientists (e.g., that all stock declines are the result of overfishing), the long-lived schism between oceanographers and fishery scientists, and the low level, until recently, of support for such research. But the fundamental explanation is the complexity of the problem which adds yet another layer of non-linearities to those already faced by the physical scientists.

Given the uncertainty of the linkages between forecasting an ENSO cold or warm event and the consequent changes in fish production, some intermediate agent is required between the physical forecasters and the fishery users of such information, a "transducer" that can take the physical forecast and analyze and interpret it for fishery managers and their customers. This intermediary might, for example, want to know for its region of interest what changes to expect in surface layer temperature and thickness, in near-surface currents (including eddies and fronts), and in special local conditions such as ice cover. This information can then be applied to an ecosystem model to yield an estimate of the likely consequences for fish production or alternative decisions on control of fishing effort.

Of course, this scientific advice is only one of the inputs, along with forecasts of economic, social, and political consequences that must be considered by the fishery managers. By this point, the precision of the distant ENSO forecast will likely have been forgotten.

REFERENCES

Baumgartner, T.R., A. Soutar, and V. Ferreira, 1992. Reconstruction of the history of Pacific sardine and northern anchovy populations over the past two millenia from sediments of the Santa Barbara basin, California. *CalCOFI*, 33, 24–40.

Beamish, R.J. and D.R. Bouillon, 1993. Pacific salmon production trends in relation to climate. *Can. J. Fish. Aquat. Sci.*, 50, 1002–16.

Francis, R.C. and S.R. Hare, 1994. Decadal-scale shifts in the large marine ecosystems of the northeast Pacific: a case for historical science. *Fish. Oceanogr.*, 3, 279–91.

Hollowed, A.B., S.R. Hare, and W.S. Wooster, 1998. Pacific basin climate variability and patterns of northeast Pacific marine fish production. *Proceedings of the 'Aha Huliko'a Workshop*, University of Hawaii, 25–29 January, 89–104.

Kawasaki, T., 1983. Why do some pelagic fish have wide fluctuations in their numbers? *FAO Fish. Rep.*, 291(3), 1065–80.

La Niña, El Niño, and US Atlantic hurricane damages

Roger A. Pielke, Jr. and Christopher W. Landsea

Recent research strongly suggests that US Atlantic hurricane damages are modulated by the phase of ENSO, with increased losses during La Niña events and reduced losses during El Niño events (Pielke and Landsea, 1998, 1999). Our analyses support the following statements:

1. La Niña means a greater frequency of damaging storms *and* more damage per storm. During cold events in the Pacific, the odds are significantly higher that the US East and Gulf Coasts will experience greater impacts because of a larger number of tropical cyclones and higher intensities for each storm.

2. Because damage increases with the square (or more) of wind speed, the greater intensity translates to a substantial increase in damage. The average damage in the United States per storm in El Niño years (Table 3-3) is $800 million vs. $1,600 million in La Niña years.

On an annual basis, because the distribution of damaging events is highly skewed by a few very large losses, we suggest that the median is an appropriate measure of central tendency. However, some decision makers with an interest in expected losses (e.g., the reinsurance industry) will be interested in the mean. Decision makers should focus on variance in losses as well as central tendency because even in a relatively inactive season, a single storm can have significant impacts. This was the case of Hurricane Andrew (1992), which resulted in more than $30 billion in losses. The largest loss in the record (normalized for inflation, population, and wealth) is the 1926 Miami hurricane which caused more than

Table 3-3 Categorization of the Atlantic hurricane season into El Niño and La Niña events

El Niño years	La Niña years
1925	**1933**
1929	**1938**
1930	**1942**
1940	1944
1941	1945
1951	1948
1953	1949
1957	**1950**
1963	1954
1965	**1955**
1969	1956
1972	1961
1976	**1964**
1977	1967
1982	**1970**
1986	1971
1987	**1973**
1990	1974
1991	**1975**
1993	1978
1994	**1988**
1997	1995

El Niño events (based upon the August-September-October Niño3.4 anomalies warmer than or equal to $+0.4\,°C$) and La Niña events (August-September-October Niño3.4 SST anomalies cooler than or equal to $-0.4\,°C$) from 1925–97. The ten most intense events of each type are highlighted in bold (Pielke and Landsea, 1999).

$60 billion in damages. This hurricane had a second landfall in the Florida Panhandle/Alabama region, which added about $10 billion in losses.
- The occurrence of an El Niño does not mean that there will be no hurricanes. Several El Niño years have seen large hurricane impacts. The 1997 hurricane season was quiet in terms of overall activity, and losses were minimal ($100 million). However, this is not always the case. In 1965 Hurricane Betsy resulted in more than $13 billion in normalized losses and in 1972 Hurricane Agnes caused more than $11 billion in damage. Thus, large losses are possible in any year, and three of the top five normalized storm losses occurred in neutral years (the remaining two were in La Niña years).
- The record suggests that Niño3.4 SSTs for the months of August–September–October provide a statistically significant indicator of damage, but the use of this relation in decision making should be with consideration of its limitations.

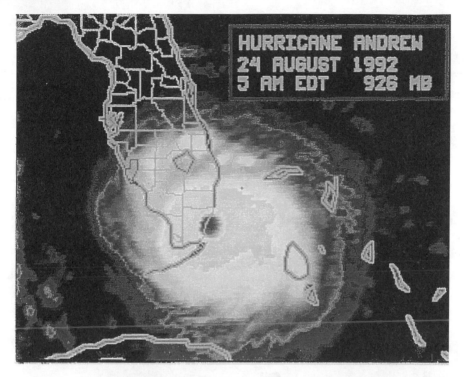

Fig. 3-11 Hurricane Andrew hitting the coast of Florida.

Common sense suggests that, with a reliable prediction of tropical sea surface temperatures in the August–September–October period, certain decision makers might be able to derive benefits. But this raises a series of questions: How reliable? Which decision makers? What benefits? Furthermore, experience offers three reasons for decision makers to exercise caution in the use of this information.

1. First, predictions are always uncertain, and a significant error in the prediction of SSTs might lead to costs rather than benefits, compared with a situation in which there is no prediction (Sarewitz et al., 2000).

2. Second, these relations, while significant, provide information with which to hedge, but should not be used to "bet an entire stake." Climate patterns change. There is always uncertainty as to how closely the future will resemble the past.

3. Third, this information will likely be of most potential value to sophisticated decision makers who can finely balance risk using probabilistic information. For an average coastal resident or community, this information might suggest accelerating preparedness plans in

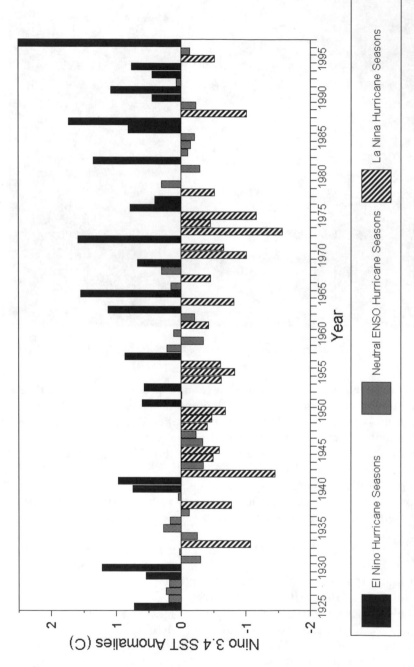

Fig. 3-12 ASO (August-September-October) Niño3.4 sea surface temperature anomalies (SSTA) for the period 1925–97. Anomalies are from a 1950–79 base period.

122

the face of a pending La Niña event, but improved preparedness also makes sense at any time.

- ENSO is not the only climate factor related to US hurricane damage, there are others that sophisticated users should consider. Other environmental factors impact Atlantic hurricanes (at least partially independent of ENSO) – such as Atlantic sea surface temperatures, the stratospheric QBO, Caribbean sea level pressures, and West African Sahel rainfall (e.g., based on the work of William Gray, 1984a,b).

About 40 percent of the years analyzed in this study (1925–95) had no significant El Niño or La Niña event occurring during the peak of the Atlantic hurricane season. Yet, substantial variations of Atlantic hurricanes and US hurricane-caused damage occur in neutral years.

A judicious use of the environmental factors, as controls in statistical models, has produced skillful experimental seasonal hurricane forecasts by Gray et al. (1993). The strong relationship between Pacific sea surface temperatures and Atlantic hurricane damages in the United States offers a tantalizing opportunity for the direct use to society's benefit of scientific information about the ENSO phenomenon. It also offers an opportunity for a closer connection between scientists and decision makers to the enrichment of both.

REFERENCES

Gray, W.M., 1984a: Atlantic seasonal hurricane frequency. Part I: El Niño and 30-mb quasi-biennial oscillation influences. *Monthly Weather Review*, 112, 1649–68.

Gray, W.M., 1984b: Atlantic seasonal hurricane frequency. Part II: Forecasting its variability. *Monthly Weather Review*, 112, 1669–83.

Gray, W.M., C.W. Landsea, P.W. Mielke Jr., and K.J. Berry, 1993: Predicting Atlantic basin seasonal tropical cyclone activity by 1 August. *Weather and Forecasting*, 8, 73–86.

Pielke Jr., R.A. and C.W. Landsea, 1998: Normalized hurricane damages in the United States, 1925–97. *Weather and Forecasting*, 13, 351–61.

Pielke Jr., R.A. and C.W. Landsea, 1999: La Niña, El Niño, and Atlantic hurricane damages in the United States. *Bulletin of the American Meteorological Society*, 80(10), 2027–33.

Sarewitz, D., R.A. Pielke Jr., and R.A. Byerly Jr., 2000: *Prediction: Decision Making and the Future of Nature*. Covelo, CA: Island Press.

Latin America

Some effects of La Niña on summer rainfall, water resources, and crops in Argentina

Guillermo J. Berri, Eduardo A. Flamenco, Liliana Spescha, Raúl A. Tanco and Rafael Hurtado

The flatlands of central and northeastern Argentina, known as the Humid Pampa, are the richest crop region of Argentina, with a total production of over 60 million metric tons (mostly soybean and corn during summer, and wheat during winter). Almost one-third of that is exported and represents a major source of income for the country. This results from the combination of good soils and appropriate weather conditions. However, most of this region is affected by the sea surface temperature variability in the tropical Pacific Ocean. Different studies have pointed out a significant relationship between ENSO (El Niño/Southern Oscillation), and seasonal to interannual climate variability in southeastern South America (see, for example, Ropelewski and Halpert, 1987, 1989). ENSO has the ability to introduce a profound modification in the general circulation of the atmosphere. In the case of central/southern South America, the alterations of the subtropical jet is the mechanism that connects the remote variability in the tropical Pacific Ocean with the local climate variability. This effect is particularly notorious in the precipitation regime around the austral (Southern Hemisphere) summer, from October through April.

Summer rainfall

In order to quantify the effect, we studied the monthly precipitation anomalies at 24 weather stations within the box defined by 57°W–63°W

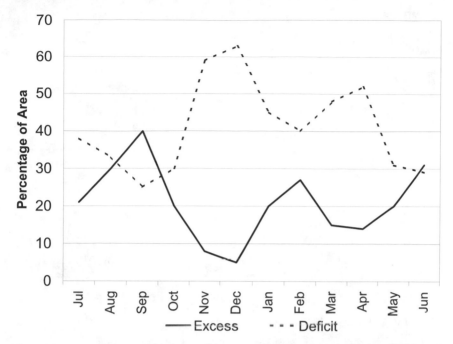

Fig. 3-13 Percentage of the Humid Pampa region (57°W–63°W, 29°S–39°S) with excess (upper tercile) and deficit (lower tercile) of rainfall. Average of seven La Niña events during 1946–93.

and 29°S–39°S (Berri and Tanco, 1998) for the period 1946–93. The original monthly values are converted in terciles, hereafter identified as below normal, normal, and above normal, and averaged over the region. The final result is expressed as a three-column matrix in which each column represents the percentage of the region that received precipitation within each tercile. Finally, two subsets are identified. One includes the years of cold ENSO events or La Niña (1950, 1955, 1964, 1970, 1973, 1975, and 1988), and the other subset includes the years of warm ENSO events or El Niño (1951, 1953, 1957, 1963, 1965, 1969, 1972, 1976, 1982, 1986, and 1991). We consider only the year when the event started. Figure 3-13, which corresponds to the La Niña composite, shows the percentage of the region that received above normal rainfall (the upper tercile) and below normal rainfall (the lower tercile). Between October and March there is a significant rainfall reduction that affects more than 60 percent of the region in December, while less than 10 percent of the region receives excess rainfall.

The rainfall shortage is more pronounced during November–December and March–April. During El Niño events (Figure 3-14), the region expe-

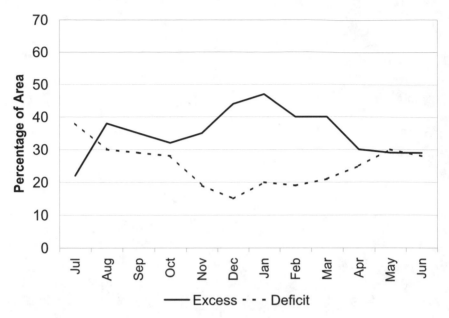

Fig. 3-14 As Figure 3-13, but with an average of 11 El Niño events during 1946–93.

riences excess rainfall in the period, to the extent that less than 20 percent of the region receives below normal rainfall. If ENSO had no effect on rainfall, all the bars should show values around 33 percent, which is the probability of occurrence of each tercile. A detailed study of six recent warm events, presented in Figure 3-15, reveals that there are always some significant dry regions (less than or equal to decile 3). This occurs despite the generalized rainfall excess experienced by the region. In contrast to that, during cold events there are no significant wet spots. It can be concluded that ENSO modifies the regional climate regime in such a way that warm events enhance rainfall, whereas cold events suppress it. The comparison of the two figures suggests that the ability of La Niña to suppress rainfall is stronger than the ability of El Niño to enhance it. It can also be concluded that the opposite extreme phases of the ENSO oscillation have a symmetric effect in the rainfall regime of this region.

Crop yields

The Humid Pampa region in Argentina, approximately located between 57°W–63°W and 29°S–39°S, is one of the richest regions of the world in terms of grain production. This results from the combination of good

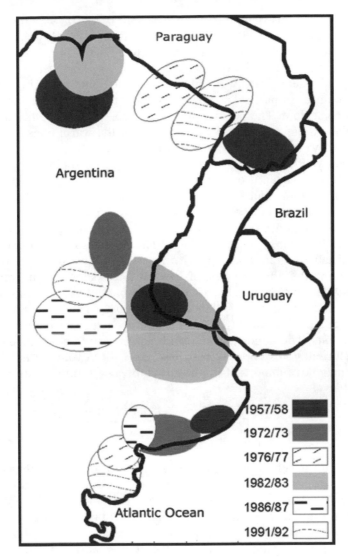

Fig. 3-15 Regions with rainfall below decile 3 during November–January of warm ENSO events initiated in 1957, 1972, 1976, 1986, and 1991. The region depicted is southeastern South America (57°W–67°W, 20°S–43°S).

soils and appropriate weather conditions. On the other hand, the region is significantly influenced by ENSO-related seasonal to interannual climate variability. Wheat production in the region accounts for 80 percent of the national production. The relationship between sea surface temperatures (SST) anomalies in the Niño3 region of the Pacific and wheat yields from 127 districts, during the period 1970–97, was investigated (Hurtado and Berri, 1998). The time series of wheat yields was detrended by means of a linear adjustment. A positive correlation coefficient between every three-month average SST during the crop cycle June to December and wheat yields was obtained for the southern part of Buenos Aires province. In particular, August–October produced the largest correlation coefficient. The positive sign of the correlation coefficient means that higher wheat yields were associated with the warm phase of ENSO, while lower yields were associated with La Niña.

Soybean yields were also studied within the core region (Spescha and Berri, 1998). Soybean is one of the most important crops in the country, totaling 16 million metric tons during the 1997–98 harvest campaign. Soybean yields of all districts in the region were correlated with the SST in the Niño3 region, averaged over every three-month period during the crop cycle November through March. The results identified two clearly distinguishable regions with opposite responses. Most of the province of Santa Fe and the northern part of Buenos Aires showed a negative correlation with Niño3, while the southeastern part of the province of Cordoba showed a positive correlation with Niño3. Again, a positive correlation with Niño3 means higher yields (lower yields) during warm (cold) ENSO events. The opposite situation holds in case of a negative correlation coefficient. This opposite response to ENSO events in neighboring regions may have important implications for agricultural planning at the regional level.

Water resources

The La Plata River Basin (1,510,000 km^2) is located in southeast South America between 26°S and 34°S. This is the second largest in the continent after the Amazon Basin and covers a great portion of Argentina and Brazil and the entire Paraguay. The main tributary is the Paraná River (annual mean flow of 12,000 m^3/s), which has a relevant importance in the region for navigation and hydroelectricity generation. The influence of ENSO on the monthly flows measured at Posadas (27°23'S, 55°53'W) during the period 1901–97 was studied by Berri et al. (2001). The analysis was similar to the one described above. The original data were converted into standardized monthly anomalies and the annual cycle was eliminated. In order to study the influence of ENSO on river flows, two data subsets were generated from the standardized anomalies of the 1901–97 period. A first group included the years of warm ENSO events, and the second group includes the years of cold ENSO events (see Table 3-4). The elements of the subsets were composites of 12 consecutive months

Table 3-4 List of warm (El Niño) and cold (La Niña) ENSO events considered in the study

Warm ENSO events or	1902/03	1905/06	1911/12	1914/15	1918/19
El Niño (27 cases)	1923/24	1925/26	1930/31	1932/33	1939/40
	1940/41	1941/42	1946/47	1951/52	1953/54
	1957/58	1963/64	1965/66	1969/70	1972/73
	1976/77	1977/78	1982/83	1986/87	1991/92
	1992/93	1994/95			
Cold ENSO events or	1904/05	1908/09	1910/11	1916/17	1924/25
La Niña (15 cases)	1928/29	1938/39	1950/51	1955/56	1964/65
	1970/71	1973/74	1975/76	1988/89	1995/96

starting in July of the year when the ENSO extreme event begins and ending in June of the following year.

Figure 3-16 shows the result of averaging the monthly river flows for the composites of El Niño years (solid line) and La Niña years (dashed line) during the 1901–97 period. It can be seen that El Niño flows are always larger than La Niña flows. The difference between the two groups peaks during November and December of the year of onset, year(0). During year(1) the difference reduces rapidly and by May of year(1) both La Niña and El Niño flows are above the average. The increased flows by El Niño always result in regional flooding during the first months of year(1). The major El Niño events (e.g. 1957–58, 1982–83, 1991–92 and 1997–98) always forced the evacuation of hundred of thousands inhabitants, some of whom remained up to six months away from their homes. On the other hand during La Niña low flows, navigation, and hydroelectricity generation is significantly adversely affected.

Berri and Flamenco (1999) presented a good example of the practical use of climate information in water resources management. The Diamante River, located in western Argentina (approximately between 34°S–35°S and 69°W–70°W), has its source in the high ranges of the central Andes Mountains. The hydrological regime presents a well-defined spring and summer maximum, when the melting of the snow that accumulated during wintertime takes place. The period of October–March accounts for 70 percent of the annual water volume. The total drainage area of the basin is 2,750 km^2. The two hydropower plants in the system, Agua del Toro and Los Reyunos, produce a combined power of 500 Mw. Important irrigation areas totaling 800 km^2, which are dedicated to vegetable crops, grapes, and other fruits, are located downstream from these water reservoirs. This represents an important economic activity in the region.

For the purpose of hydroelectricity management, the accumulated water volume flowing during the period October–March is used as a measure of the water available in the system. At the end of the Southern

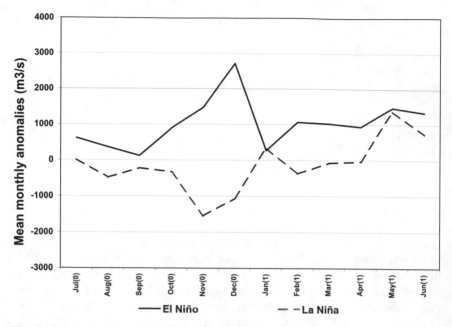

Fig. 3-16 Standardized anomalies of monthly river flows of the Paraná River at Posadas, averaged for the composite of El Niño years (full line), La Niña years (dashed line) during the period 1901–97. The index (0) indicates the year when the event begins and (1) the year when the event ends.

Hemisphere winter (in September), an estimate of the volume of the snow deposited in the catchment is made and the first seasonal volume prediction is issued. The model in use converts the snow volume accumulated in the catchment into an equivalent water volume that will flow during the period October–March. The power utility company uses this prediction to produce future electricity generation estimates. A statistically significant positive correlation was found between the seasonal volume (October–March) and sea surface temperatures (SSTs) in the Niño3 region of the Pacific Ocean (5°S–5°N, 90°W–150°W), during March–April and November–December. A multiple linear regression model for the October–March volume predictions was developed, making use of Niño3 SST anomalies observed during March–April and November–December. A validation analysis was performed for the period 1981–94, developing the model coefficients for the training period 1949–80. Since the November–December Niño3 SST anomalies were input to the model, they were replaced with six-month Cane-Zebiak model predictions (Chen et al., 1995; Zebiak and Cane, 1987), performed in May.

A contingency analysis of the three-category forecasts, i.e., terciles

Table 3-5 Cross-validation analysis

	Predicted		
	Below	Normal	Above
Observed			
Below	2	2	0
Normal	2	3	0
Above	0	0	5

Cross-validation analysis between observed October–March volume anomalies in the Diamante River (Argentina) and predicted volume anomalies with a regression model based on observed Niño3 SST during the previous March and April and six-month predictions of simultaneous November and December Niño3 SST. The training period is 1949–94 and the cross-validation period is 1981–94.

(shown in Table 3-5) produced a categorical skill of 71 percent, which means 10 out 14 correct forecasts. Although a sample of 14 cases is not large enough to draw definitive conclusions about the model, we consider the results sufficiently positive to continue to search for other relationships in order to improve the skill of the method. The advantage of the model based on the SSTs resides in its ability to produce a forecast in May, with similar results to the traditional snowpack thickness model in use, even before the main snowfalls. The snow cover model can only be applied in September when the maximum snow cover is reached. On the other hand, the volume of snow deposited in the catchment is estimated from a few point measurements and, therefore, it is only approximated. With this model, the water resources operator has the advantage of having information about the amount of water available in the system four months earlier.

The 1998–2000 La Niña

Regarding rainfall during the summer semester 1998–99, most of southern Brazil, central and eastern Paraguay, northeastern Argentina and western Uruguay experienced seasonal rainfall below 75 percent of the average. Large portions of the regions also received less than 50 percent, particularly during October–December 1998 and January–March 1999 over southeastern Paraguay, northeastern Argentina, and south-central Brazil. During 1999, Paraguay continued to experience below normal rainfall which resulted in one of its most severe droughts in the last 20 years. During the 1999–2000 summer semester, the core of the dry region moved towards the extreme south of Brazil, most of Uruguay and central eastern Argentina. Again, large portions of this territory received less than 50 percent of the average rainfall for the season. During the 1999–2000 summer, Paraguay continued to experience dry conditions. This general behavior was in agreement with the results obtained in different studies that indicated generalized below normal rainfall during the period October–March in southeast South America.

Regarding water resources during the 1998–99 summer period, the Paraná River did not depart significantly from the average, in contrast with the 1999–2000 period, when it experienced one of its lowest levels. In this last period the conditions in the basin were in accordance with the averaged behavior displayed by the composite of 15 La Niña events since 1901. The low levels registered during the summer period had a negative effect on the navigation of the system, particularly in the other main tributary of the La Plata River Basin, the Paraguay River. The shipment of grains via the river registered an important increase, due to the reduced barge draught.

Regarding crops, the 1998–99 agricultural campaign did not really suffer from the moderately dry conditions in the core of the crop region of Argentina (northwest of Buenos Aires, south of Santa Fe and southeast of Córdoba provinces). Preliminary reports indicated that corn yields in the core region would be low, as would soybean in the north and northeast of Buenos Aires province and in central and southern Santa Fe province.

ACKNOWLEDGMENT

This research was partially supported by Research Grant PIP-0389/98 of the National Research Council of Argentina (CONICET).

REFERENCES

Berri, G.J. and E. Flamenco, 1999: Seasonal volume forecast of the Diamante River, Argentina, based on El Niño observations and predictions. *Water Resources Research*, 35(12), 3803–10.

Berri, G.J. and R. Tanco, 1998: Some effects of El Niño in the summer rainfall in the Humid Pampa of Argentina. *X° Brazilian Congress of Meteorology*, Brasilia, October 1998.

Berri, G.J., M.A. Ghietto, and N.O. García, 2001: The influence of ENSO in the flows of the Upper Paraná River of South America over the past 100 years. Accepted for publication by the *Journal of Hydrometeorology*.

Chen, D., S.E. Zebiak, A.J. Busalacchi, and M.A. Cane, 1995: An improved procedure for El Niño forecasting: Implications for predictability. *Science*, 269, 1699–702.

Hurtado, R. and G.J. Berri, 1998: Relationship between wheat yields in the Humid Pampa of Argentina and ENSO, during the period 1970–1997. *X° Brazilian Congress of Meteorology*, Brasilia, October 1998.

Ropelewski, C.F. and M.S. Halpert, 1987: Global and regional scale precipitation patterns associated with the Niño/Southern Oscillation. *Monthly Weather Review*, 115, 1606–26.

Ropelewski, C.F. and M.S. Halpert, 1989: Precipitation patterns associated with high index phase of Southern Oscillation. *Journal of Climate*, 2, 268–84.

Spescha, L. and G.J. Berri, 1998: On the effect of ENSO on the soybean yields in the Humid Pampa of Argentina. *X° Brazilian Congress of Meteorology*, Brasilia, October 1998.

Zebiak, S.E. and M.A. Cane, 1987: A model of El Niño/Southern Oscillation. *Monthly Weather Review*, 115, 2262–78.

La Niña effects in Ecuador

M. Pilar Cornejo-Grunauer

The probable occurrence of a La Niña event was announced in Ecuador by March 1998 (*El Universo*, 1998). After June 1998, when the 1997–98 El Niño had disappeared, there was a brief period of demand for information about La Niña's impacts. The general public had not heard about it before, and the damage caused by El Niño was reason enough to be ready or at least informed about another extreme climate-related event.

By the end of December 1998 the demand for La Niña information was coming mostly from the agriculture sector. This was also related to Hurricane Mitch (October 1998), which destroyed Ecuador's competitors in agricultural exports, especially in bananas, among other export commodities like melon, asparagus, and onion. When Mitch devastated agricultural production in several Central American countries, it opened a window of opportunity for these exports from Ecuador.

Current and common knowledge suggested that La Niña was the opposite of El Niño, but available scientific literature suggests the following:

- There is not much agreement about the cold extreme of the ENSO cycle. The most popular name is La Niña, although some people still use El Viejo (the old one) or anti-El Niño, or the cold phase.
- There is no universally accepted formal definition of La Niña. What variables would be used to define it? Sea surface temperatures (SSTs)? Surface winds? How big do the SSTs or surface winds anomalies have to be? How long do the anomalies have to last?, etc.

In spite of the lack of a formal definition, the following are listed as La Niña signature conditions off the Ecuadorian coast:

- Oceanic temperature fields drop 2°–3 °C below normal (i.e., average) for periods longer than the usual dry season (May through November) (Sonnenholzner and Cornejo-Rodríguez, 1995).
- The thermocline (defined by the 20 °C isotherm) surfaces near the coast.
- Upwelling Kelvin waves are more frequent.
- The 13 °C isotherm is stronger.
- Depending on its timing, it has a negative effect on precipitation, suppressing rainfall for a month (usually March) or decreasing it during the rainy season (December through April) or enhancing the dry season (May through November) along the coastal plains.

La Niña impacts

Some analyses of climate and socio-economic information show that there is an impact on health, fisheries and aquaculture, and agriculture sectors. In the past, La Niña's impacts on the fisheries of small pelagic fish such as sardines (*Sardinops sagax*) and others used for fish meal were positive, but during the 1998–99 La Niña, these stocks had not recovered yet from the impacts of the 1997–98 warm event, which caused these pelagic stocks to migrate southward.

La Niña's impacts on shrimp aquaculture and fisheries are negative: the availability of wild shrimp larvae decreases and the shrimp population migrates to the Panama area. The first problem (the decrease in wild shrimp larvae) was solved after the 1985 La Niña, with the creation of shrimp hatcheries in Ecuador, where gravid female shrimps are used for spawning, or where shrimp maturation takes place. This provides enough "seed" stock for the sector (Cornejo-Grunauer, 1998). Knowing there was a direct relationship between ocean temperatures and shrimp population (Figure 3-17), the Niño3 index and the price of shrimp larvae from the hatcheries were used to develop a forecast model for shrimp larvae availability. Since 1996, these modeling results were provided to the sector in order for it to pursue mitigation and adaptation measurements (Cornejo-Grunauer et al., 1997). However, the scenarios were different for the 1998–99 La Niña:

1. The seasonal low in shrimp larvae availability during the 1999 dry season was enhanced by cooler than normal conditions.
2. In March 1999, there was an epidemic outbreak of a shrimp disease in Central America, which was then followed by the White Spot Syn-

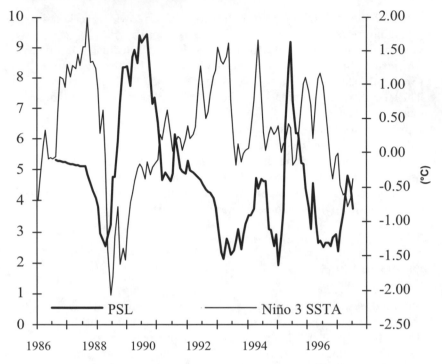

Fig. 3-17 Price of shrimp larvae (PSL) inverted to represent shrimp larvae availability and El Niño3 SST anomalies.

drome virus (WSSV), which spread southward affecting shrimp production in Ecuador's hatcheries and ponds. So there is no way now to measure what part of the losses in the sector resulted from the 1998–99 La Niña and what part resulted from the WSSV.

Effects in agriculture were mainly the result of changes in precipitation patterns: the starting date of the rainy season, the duration of the rainy and dry seasons, and the amount of solar radiation and cloud cover. This sector is much more familiar with El Niño's impacts than with La Niña's impacts. Figure 3-18 (c) shows the yield for two crops at Quevedo (approximately 1°6′S, 79°27′42″), rice during the rainy season and soy during the dry season with the time series sea surface temperature anomalies for the Niño3 region. The data for the rice yield for 2000 are estimated. One can see the effects of La Niña for the years of 1985, 1988–89, 1996 and 1998–99, but the effects are mixed. Thus, one has to be very careful about attribution of an impact to an ENSO extreme. Also, each La Niña event has been different and, here, we defined its magnitude in terms of its sea surface temperature anomalies and their duration: 1985 – moderate; 1988–89 – strong; 1996 – weak; 1998–99 – weak. On the other hand, its

Table 3-6 Total amount of rain for the rainy and dry seasons in Pichilingue (1°6'S, 79°27'42''), near Quevedo, Ecuador

La Niña year	Rainy season Total (mm)	%	Rice yield	Dry season Total (mm)	%	Soy yield
Mean (1981–98)	1890.18			527.18		
total 1985 moderate	1136.7	60	−0.13	120.3	23	1.71
total 1988 strong	1388	73	−0.22	258.9	49	−0.07
total 1989 strong	2015.2	107	−0.72	216.8	41	−0.31
total 1996 weak	1398.9	74	0.57	52.9	10	−0.98
total 1998 weak	3528.2	187	−1.92	1220.20	231	−0.91
Total 1998 Sep–Nov				92	40	
Total 1999			−0.59			0.15

The rainy season extends from December through April, and the dry season from May through November. During 1998, only the end of the dry season can be considered as part of a La Niña episode. Rice and soy yields normalized time series values are shown (source: INAMHI).

effects also depend on the geographical location within the country. The location shown in Figure 3-18 is near the foothills of the Andes. Therefore, depending on the year, its rainfall can be enhanced or decrease due to El Niño and La Niña. During a strong El Niño event, rainfall is enhanced all over the coast. During moderate to strong La Niña events, rainfall is enhanced in the foothills of the Andes with a similar La Niña impact in neighboring Colombia. Table 3-6 shows the total rainfall for the rainy season (December–April), and the dry season (May–November) and their percentages with respect to the mean total rainfall for each season. During 1985, 1988–89 and 1999, La Niña's impacts on rice production were negative, because it rained less than the rainy season mean. The rainy season of 1998 was similar to that of the 1997–98 warm event. For soy, the impacts of the 1988–89 and 1996 La Niña events were are also negative. However, for both crops there were also positive impacts during 1996 for rice and during 1985 for soy. It is important to mention that the owner of "Aguas Claras" (Mr. Duque, personal communication) did not plant soy during the 1997 dry season. Having been provided with information about El Niño's occurrence, he believed it. During December 1999, when some rainfall appeared, producers decided to plant rice ahead of time without taking into account information about La Niña, which at the time did not receive the same level of attention as El Niño. Later on, during January and February 2000, it rained less than normal and there were crop losses. They usually planted according to their personal farming experience with regard to when the onset of the rainy season would start and did not base their activities on the use of any other climate information.

There are some other effects in the agriculture sector, but they have not yet been quantified. For example, during La Niña air temperature is lower, and there might be more cloudiness. As a result, bananas still on the plants might not be able to attain the "grading" they require for export. Other plants, such as mangoes, might not flower.

Unquantified effects were found in tourism and health as well. Tourism at beach resorts decreased, because air and water temperatures were cooler than normal and not suitable for water sports. Other impacts might be positive, e.g., a national clothes-manufacturing industry would sell more sweaters and light jackets than in a regular dry season, due to the wind factor.

In general, there is a need for increased research on La Niña causes, manifestations and its impacts. If one believes that there is enough evidence to support interdecadal and multidecadal variability as long-term time series show, one has to be ready for the next "cold phase" of this interdecadal oscillation where we should expect more La Niñas than El Niños, as happened between 1950 and 1975. At ESPOL and within the Trade Climate Convergence Complex (TC3; see the website at www2.usma.ac.pa/~cathalac/tccc.htm), researchers have assumed the task of understanding ENSO's extreme events and their relationship to important socio-economic sectors, not only in Ecuador but in the TC3 region as well.

REFERENCES

Cornejo-Grunauer, M.P., 1998: Variaciones climáticas: Impacto del Fenómeno de El Niño/Oscilación Sur en la Acuicultura de Ecuador, *Boletin CENAIM*.

Cornejo-Grunauer, M.P., J. Calderón, J.L. Santos, and G. Silva, 1997: Application of climate information in shrimp aquaculture: the Ecuadorian case. *ENSO Signal*, IRI-OGP-NOAA, September.

El Universo, 1998: Report on ESPOL and PMRC work in the newspaper edition of 5 March.

Sonnenholzner, S. and M.P. Cornejo-Rodríguez, 1995: Subsurface variability associated with the Cromwell current between the Ecuadorian coast and 82°W, *Tecnologica-ESPOL*.

La Niña's impacts in Cuba:
The opposite side of the coin?

Lino Naranjo-Diaz

La Niña, or more properly, the cold phase of the ENSO cycle, has to be considered as part of a seesaw-like pattern, where El Niño represents its most dramatic extreme. El Niño has monopolized the interest of the scientific community dealing with the forcing elements responsible for interannual climatic variability.

Glantz (1996) has attempted to explain this condition. He has argued that, because cold events produce adverse weather and climatic anomalies that often *seem* to be opposite to those produced by El Niño, cold events are perceived by many to be periods when weather and climate conditions are considered to have returned to normal. However, there still remains a lot of uncertainty about La Niña. Is La Niña really the "good part" of the ENSO cycle to some countries? How "normal" could the climate be in regions under La Niña's influence? In addition, the list of El Niño and La Niña events after 1950 (Trenberth, 1997) shows a lower number of La Niña events than El Niño (15 vs 10). After 1972, the number of El Niño events has been higher than those of La Niña occurrences, suggesting a "warming" tendency in the ENSO cycle over the last three decades of the twentieth century. All these facts reinforce the main question: Is La Niña information, including forecasts, "usable science?" Should La Niña events be highlighted as having important impacts on society, or are they only of marginal importance? Answers will likely be very different, depending on the region of the globe under consideration. In this case, some facts and uncertainties about the impacts of La Niña on

139

Fig. 3-19 Geographical location of Cuba.

the island nation of Cuba are analyzed in an attempt to address some of these questions.

La Niña's impact on the climate and economy of Cuba

The Republic of Cuba is an independent state, located in the Meso-American Caribbean Sea at the entrance to the Gulf of Mexico (Figure 3-19). It is the biggest island state in this basin and it is an archipelago of 110,860 km² in area, of which 105,599 km² is the main island (Cuba). The remainder includes Juventud Island and 1,600 other islands and keys.

The characteristics of Cuba's climate are conditioned by the location of the archipelago at the northern edge of the tropical area, and by other features like the elongated shape of the main island (Cuba) with 5,746 kilometers of coastline. Being in the vicinity of North America as well as important atmospheric and oceanic systems such as the trade winds and the Gulf Stream, respectively, also influences the Cuban climate which can be considered tropical or tropical maritime. It is characterized by a well-defined rainy season (also called "wet season") with warm temperatures, and a dry season when precipitation is more scarce.

As in many countries around the world, Cuban studies about the ENSO cycle have been closely related to and focused on the more dramatic El Niño. Cardenas and Naranjo (1996) studied the influence of ENSO events on the climate of Cuba, including both warm and cold events. Their study, undertaken in the Cuban Climate Center, used a chronology of ENSO

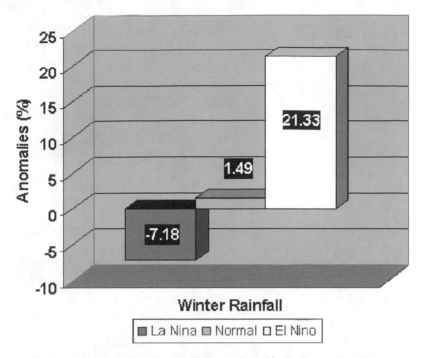

Fig. 3-20 Rainfall anomalies (percent) for Cuban winter during La Niña, normal, and El Niño conditions in the tropical Pacific.

events based on a non-conventional "ENSO Index" (IE in Spanish) which combines the Southern Oscillation Index (SOI) and sea surface temperature anomalies in the Niño3 region, as follows:

$$\text{ENSO Index (IE)} = -\text{MSSTA (MSOI) for MSOI} < 0$$

$$\text{IE} = \text{MSSTA (MSOI) for MSOI} > 0$$

where MSSTA and MSOI are the three-month running means, respectively, for the sea surface temperature anomalies in Niño3 and the SOI.

Like El Niño, the influence of La Niña over the climate elements in Cuba seems to be most significant in the Northern Hemisphere winter. However, its influence is, to some extent, the opposite of El Niño's influence. With respect to rainfall, for instance, El Niño is associated with a significant increase in winter rainfall, which consequently defines positive anomalies in monthly amounts. Under La Niña conditions, such anomalies are less significant and tend toward a slight deficit in rainfall (Figure 3-20). However, there is considerable variability in this behavior. Carde-

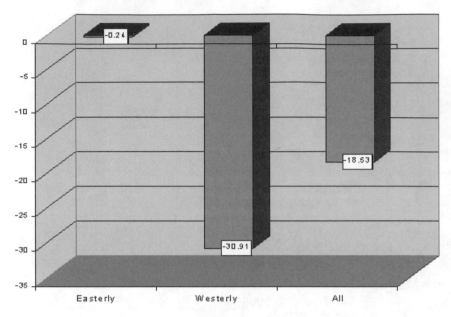

Fig. 3-21 Rainfall anomalies (percent) in QBO phases.

nas and Naranjo (1997) established the idea that some other factors, which they called "modulating factors," tend to modify the expected impacts of ENSO's extreme phases on climate elements in Cuba. Such factors explain a significant fraction of this variability. Figure 3-21 shows the influence of the Quasi-Biennal Oscillation (QBO) in the lower stratosphere acting as a modulating factor on La Niña's impact on rainfall in the western region of Cuba. This oscillation represents a regular seesaw-like pattern with easterly and westerly phases in the zonal (east-west) winds over the tropical belt, and this result seems to indicate a significant dryness in this region of Cuba, when La Niña's influence is coincident with the westerly phase of the QBO. During the 1998–99 La Niña, for instance, an important rainfall deficit was observed in winter. This event followed the intense drought in summer 1998, which has been associated with the 1997–98 El Niño, producing a long-lasting drought that had catastrophic effects on the Cuban economy. Occasionally, La Niña's winters have been associated with normal effects. However, it is important to note that it is difficult to identify reliable attributions to ENSO's impacts.

Regarding temperatures, significant negative anomalies are associated with La Niña, whereas during El Niño there is not a clear significant impact. This is one of the very few instances where the La Niña signal over a Cuban climatic element is stronger than during El Niño.

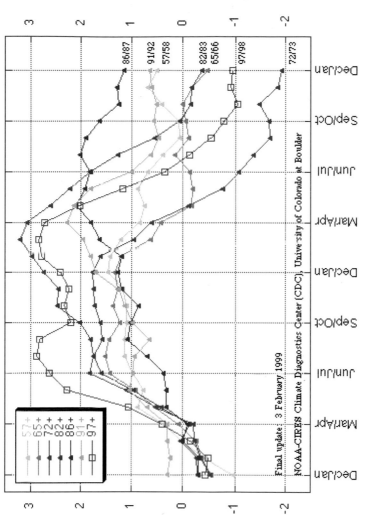

Fig. 1-1 Multivariate ENSO Index for the seven strongest historic El Niño events since 1950. El Niño/ Southern Oscillation (ENSO) is the most important coupled ocean-atmosphere phenomenon to cause global climate variability on interannual time scales. Here an attempt is made to monitor ENSO by developing the Multivariate ENSO Index (MEI) based on the six main observed variables over the tropical Pacific. These six variables are: sea level pressure (P), Zonal (U) and meridional (V) components of the surface wind, sea surface temperature (S), surface air temperature (A), and total cloudiness fraction of the sky (C) (from www.cdc.noaa.gov/~kew/MEI/).

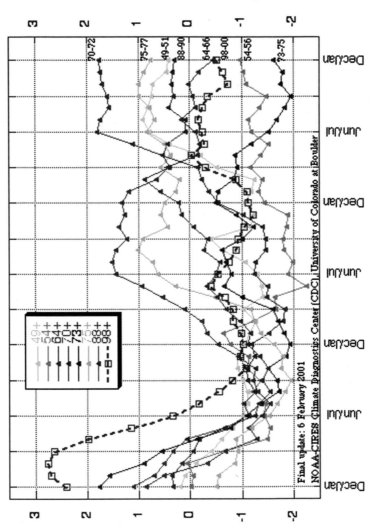

Fig. 1-2 Multivariate ENSO Index for the eight strongest historic La Niña events since 1949 vs recent conditions (from www.cdcd.noaa.gov/~kew/MEI/).

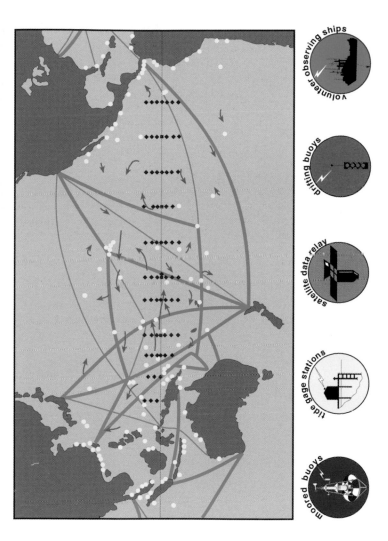

Fig. 1-13 The El Niño/Southern Oscillation (ENSO) Observing System was set up to understand, monitor, and predict year-to-year climate variations in the tropical Pacific associated with El Niño and La Niña. The ocean-based components, shown here, relay data to shore in real time via satellites. The main components are a volunteer ship program (blue lines), an island and coastal tide-gauge network (yellow circles), a system of drifting buoys (orange arrows), and the Tropical Atmosphere Ocean (TAO) array of moored buoys (red diamonds and squares). These measurement platforms provide data for surface winds, sea surface temperature, upper ocean thermal structure, sea level, and ocean currents. Complementing this ocean-based network are satellites that provide data from space with near-global coverage (From NOAA/PMEL/TAO Project Office, www.pmel.noaa.gov/tao/proj_over/diagrams/).

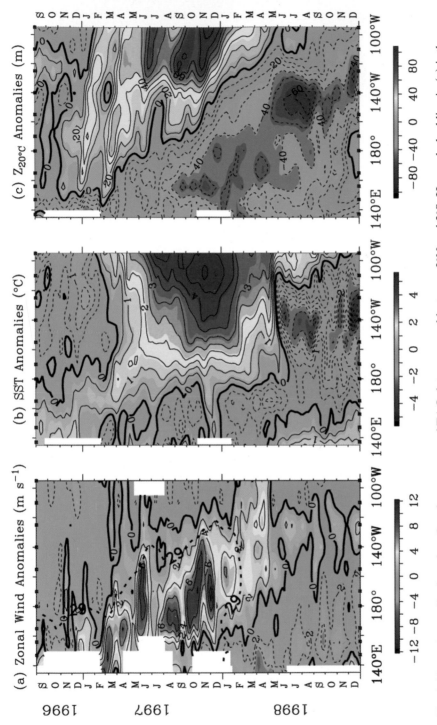

Fig. 1-14a–c Anomalies based on five-day averages of TAO data averaged between 2°N and 2°S. Heavy dashed line in (a) is the 29°C isotherm through early 1998. The depth of the 20°C isotherm in (c) is a commonly used indicator for thermocline depth. Little boxes at top and bottom show data availability.

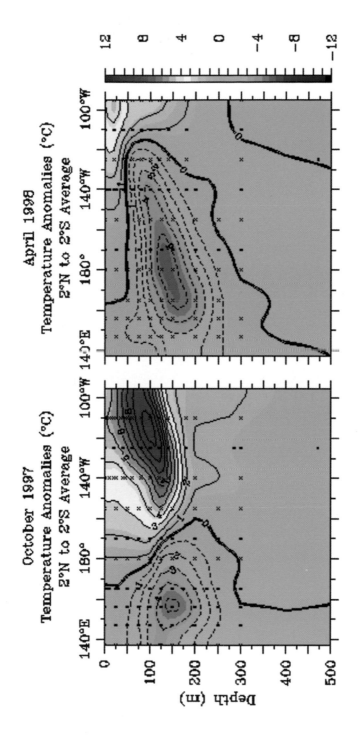

Fig. 1-16a–b Monthly mean temperature anomalies (°C) from the TAO array averaged between 2°N and 2°S for (a) October 1997 and (b) April 1998. The small x symbols mark longitudes and depths of data availability.

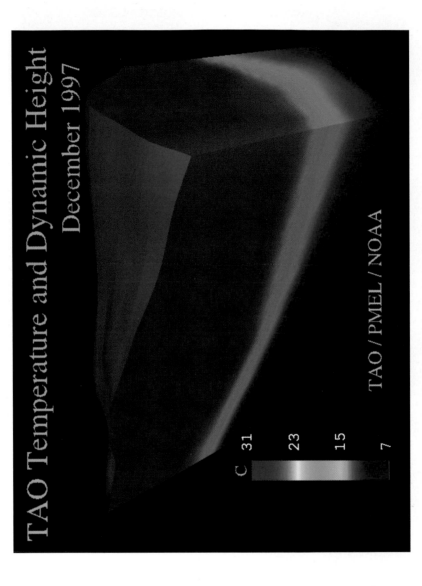

Fig. 2-12 December 1997 temperature (from NOAA/PMEL/TAO Project Office, www.pmel. noaa.gov/tao/proj_over/diagrams/).

COLD EPISODE RELATIONSHIPS   DECEMBER - FEBRUARY

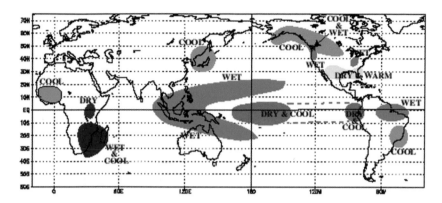

COLD EPISODE RELATIONSHIPS   JUNE - AUGUST

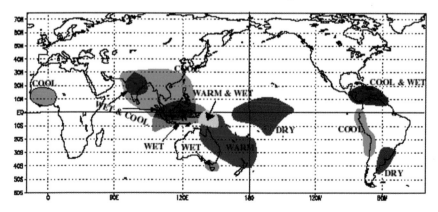

Fig. 2-2 Cold episode relationships, December–February and June–August (from www.cpc.ncep.noaa.gov/products/analysis_monitoring/laNiña/cold_impacts.html).

WARM EPISODE RELATIONSHIPS   DECEMBER - FEBRUARY

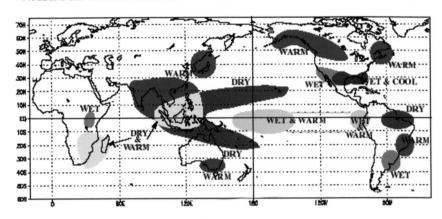

WARM EPISODE RELATIONSHIPS   JUNE - AUGUST

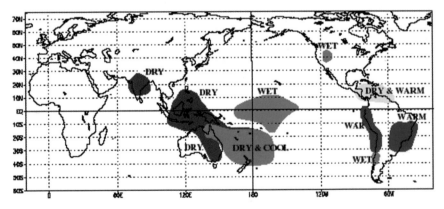

Fig. 2-3 Warm episode relationships, December–February and June–August (from www.cpc.ncep.noaa.gov/products/analysis_monitoring/impacts/warm_impacts.html).

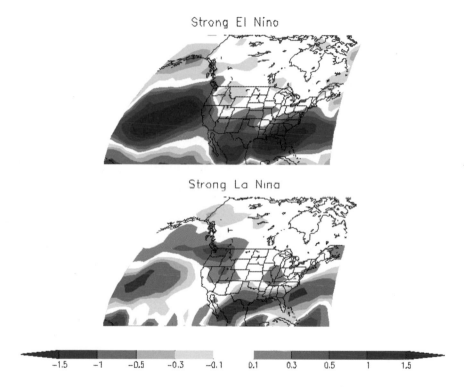

Fig. 2-5 General circulation model simulations of North American wintertime precipitation anomalies for an extreme El Niño case (top) and an equally strong extreme La Niña case (bottom). Results are based on a 10-run ensemble average. The strong warm case mimics tropical Pacific sea surface temperature anomalies of 1982–83, and these anomalies are simply flipped in sign to generate the strong cold case. Positive (wet) anomalies are blue, and negative (dry) anomalies are red.

Fig. 3-4 Map indicating the three-state Pacific Northwest region: Idaho (ID), Oregon (OR), Washington (WA), and the Columbia River Basin (outlined in orange), that is the focus of this study. Note that a significant portion of the Columbia River Basin resides in Canada, while relatively small portions reside in western Montana (MT), Wyoming (WY), and Nevada (NV).

Strong La Niña year October–March Precipitation anomalies

Strong La Niña year October–March Surface temperature anomalies

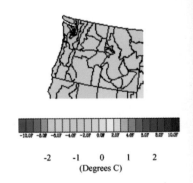

Fig. 3-7 Composite October–March precipitation and temperature anomalies for the strong La Niña years identified in Table 3-1. Images were obtained from NOAA's Climate Diagnostic Center Website at www.cdc.noaa.gov/Usclimate/Usclimdivs.html.

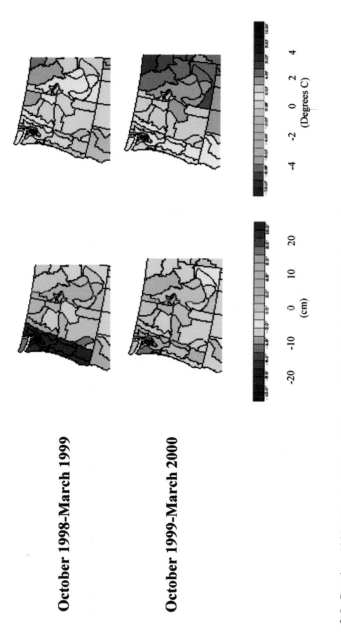

Fig. 3-9 October 1998–March 1999 and October 1999–March 2000 US climate division surface temperature and precipitation anomalies, relative to the 1950–95 climatology (from NOAA's Climate Prediction Center).

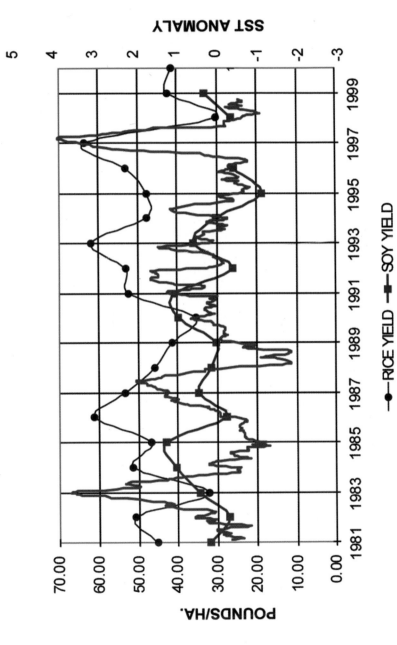

Fig. 3-18 Rice and soy yields from the Aguas Claras farm near Quevedo, Ecuador (approximately 1°6′S, 79°27′42″). The gray line is El Niño3 SST anomalies for the same period.

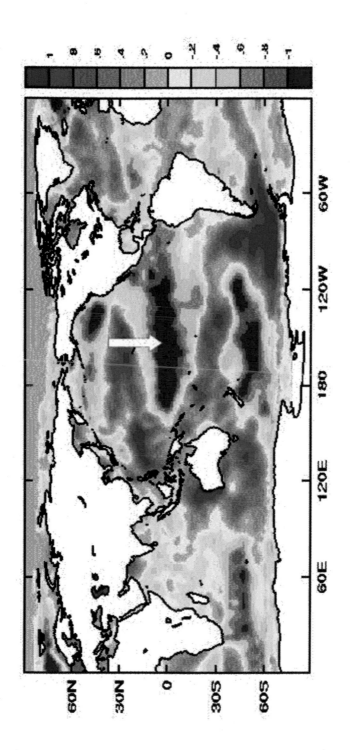

Fig. 3-29 SST anomaly analysis of the Western Pacific during a La Niña event. Note the characteristic cold tongue (arrow) extending from the eastern Pacific to the western Pacific (from Allan et al., 1996).

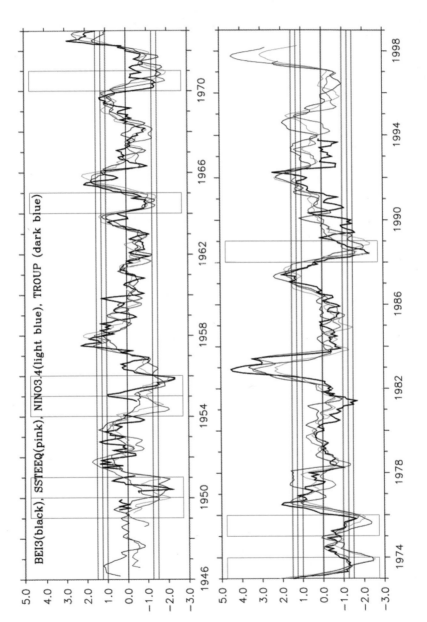

Fig. 5-1 Overlay of several time-series indices to define the existence of cold events in the tropical Pacific Ocean.

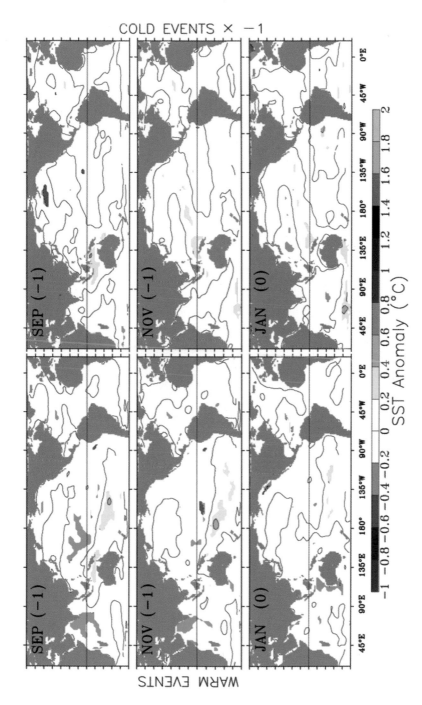

Fig. 5-2a–d The composite (average) SST anomaly patterns associated with the nine cold events and the ten warm events between 1946 and 1993.

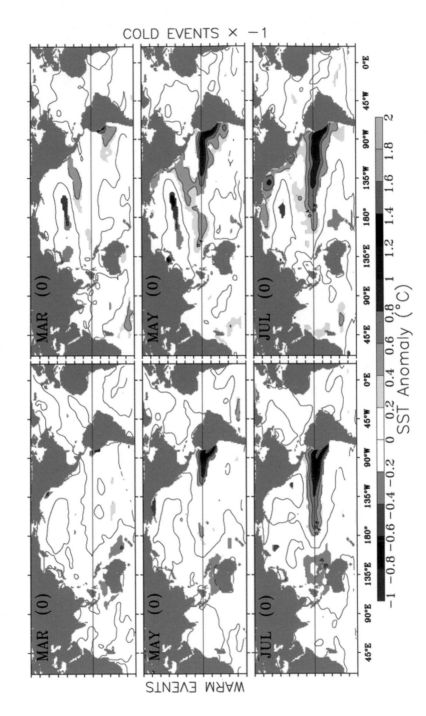

Fig. 5-2a–d The composite (average) SST anomaly patterns associated with the nine cold events and the ten warm events between 1946 and 1993.

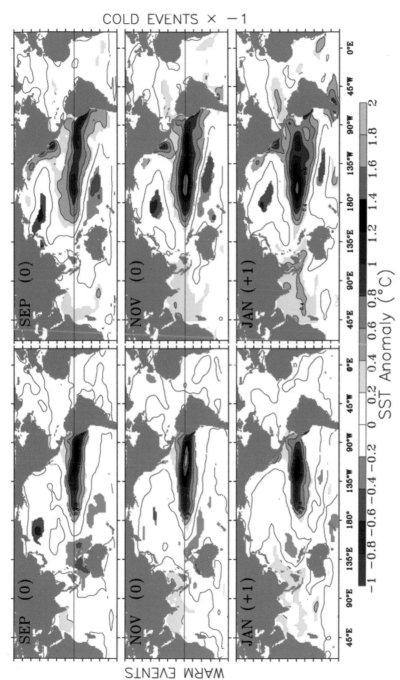

Fig. 5-2a–d The composite (average) SST anomaly patterns associated with the nine cold events and the ten warm events between 1946 and 1993.

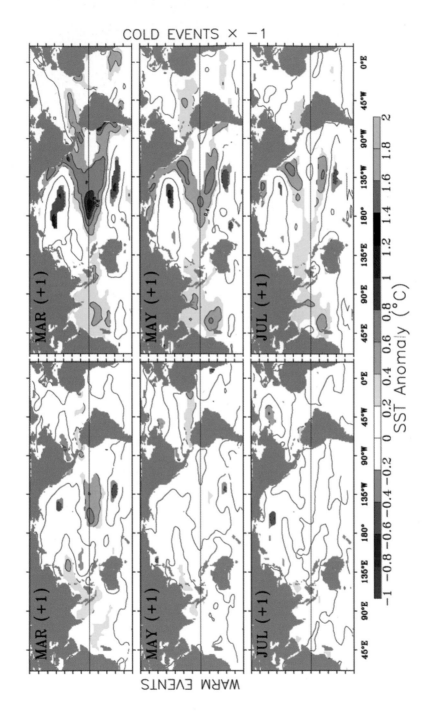

Fig. 5-2a–d The composite (average) SST anomaly patterns associated with the nine cold events and the ten warm events between 1946 and 1993.

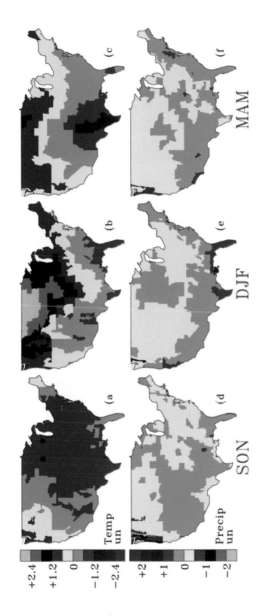

Fig. 5-3a–c A variety of results pertaining to the seasonal temperature and precipitation anomalies that have occurred during the nine cold event years since 1946.

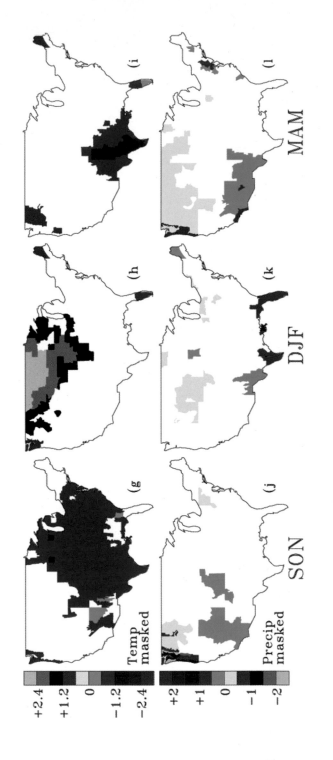

Fig. 5-3a–c A variety of results pertaining to the seasonal temperature and precipitation anomalies that have occurred during the nine cold event years since 1946.

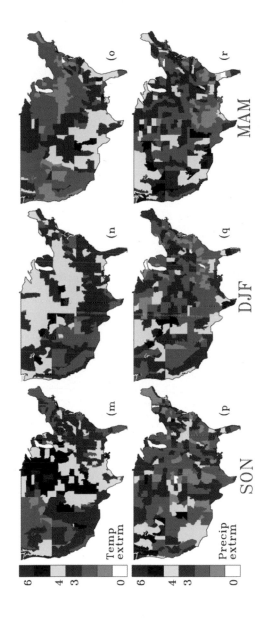

Fig. 5-3a–c A variety of results pertaining to the seasonal temperature and precipitation anomalies that have occurred during the nine cold event years since 1946.

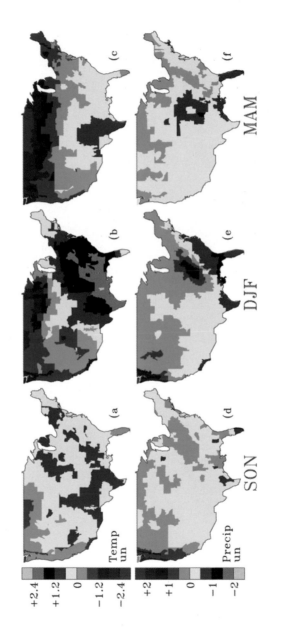

Fig. 5-4a–c The same types of seasonal weather anomaly information as Figure 5-3, but for warm event seasons.

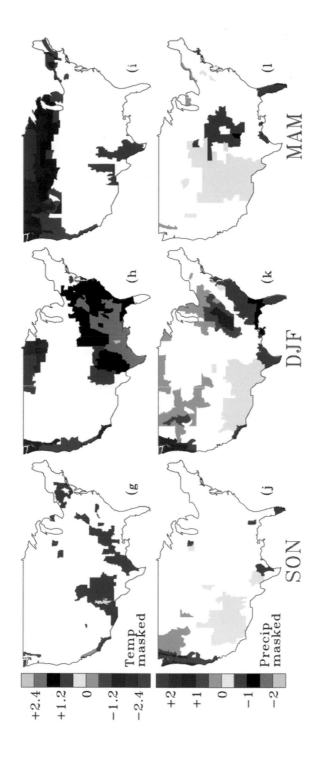

Fig. 5-4a–c The same types of seasonal weather anomaly information as Figure 5-3, but for warm event seasons.

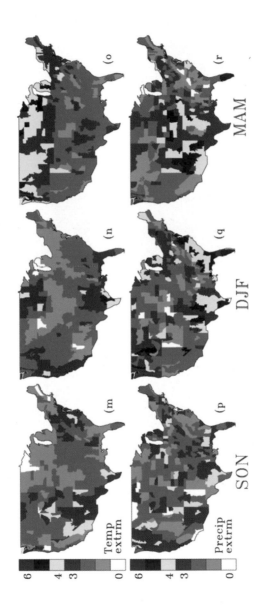

Fig. 5-4a–c The same types of seasonal weather anomaly information as Figure 5-3, but for warm event seasons.

Regardless of the specific modulating factors that can affect El Niño's impacts in Cuba one way or another, one might conclude that during cold events, Cuban winters tend to be slightly cooler with near-normal to slightly below-normal rainfall and that there is no significant influence in the summer. However, one must ask – how reliable are these statements? The former conclusions were based on an analysis of monthly mean values in which several important statistical features *were not* considered. Regarding the influence of La Niña on Cuban weather and climate in the summer, some factors support the idea that a more significant influence could exist. For instance, during the strong La Niña of 1998, anomalously heavy rainfall was recorded over the central region of Cuba, producing catastrophic effects, mainly in Cienfuegos Province.

Another interesting point about the contradictory nature of La Niña's impact on climate is its relationship with tropical cyclone activity. Gray (1984) suggested the existence of significant relationships between the anomalies of sea surface temperatures in the tropical Pacific Ocean and tropical cyclone activity in the Atlantic, where he found a tendency toward fewer cyclones in El Niño years (see also Pielke and Landsea, this volume). Consequently, Gray's finding suggests that the anomalies imposed on the summer circulation in the Atlantic during an El Niño event tend to reduce the efficiency of tropical cyclones to intensify to hurricane status in the Caribbean and Atlantic basins. During a cold event, however, the opposite trend is to be expected. Therefore, under La Niña conditions, tropical storms and hurricane activity in the Atlantic basin are likely to increase, as opposed to the situation during an El Niño event. Consequently, one might expect that the risk of a tropical storm hitting Cuban territory increases during a La Niña episode. However, this expectation is in contrast with the fact that the frequency of tropical storms that affected Cuba during cold events in the 1950–98 period was the same as the long-term climatic frequency. Consequently, tropical storm behavior is close to what we consider as normal.

Regarding the impacts on some economic activities, attributions become more complex, because our knowledge about the mechanisms linking La Niña to impacts is in its early stages and some relationships are hard to identify with confidence. Naranjo (1997) pointed out the influence of ENSO on some key winter crops in Cuba. As one example, potato crop production in the 1995–96 winter (during a La Niña episode) nearly doubled in the 1997–98 winter (during an El Niño episode) and was the highest in the decade, suggesting an impact on this crop which is very sensitive to rainfall and temperatures anomalies in winter. A cooler and dryer climate in winter is very favorable for potato yield. However, for sugar cane there are no clear signs of impacts during the same period. Sugar content in plants is stimulated by dry conditions in the winter (the harvest season), but changes induced by La Niña seem not to be enough to significantly improve yields. However, dryness could have a negative feedback effect on sugar cane because, at the same time, young plants could be severely affected, thereby reducing the estimated yields for the harvest of the next year.

Additionally, there is another factor that should be taken into account when La Niña's impact on the Cuban economy is analyzed. From February to the middle of May, the frequency of occurrence of very intense pre-frontal squall lines with associated tornadoes and hail is significantly increased in El Niño years with heavy damages in agriculture. During La Niña, these events are practically absent and, from this point of view, La Niña impacts could be considered as "positive." Yet, again in this case there is no significant difference between La Niña and normal years.

Ortiz and Guevara (1997) found an ENSO signal in Cuba in the interannual variability of the disease *Meningitis meningococica*, although this signal is, for the most part, related to a significant increase in the disease during El Niño years. Under La Niña conditions, it is nearly normal.

Conclusion

It is clear from the above discussion that there are important examples of where La Niña episodes seem to behave like "normal years." However, any conclusions about these examples should be made with caution, because some of them could be dependent on methods for determining the attribution. One important reason why methods used to isolate an El Niño signal do not work for La Niña is that many of our time series (for agriculture, health, etc.) begin after 1970, and, as noted earlier, this period was conditioned by an increase in the frequency of warm events and a decrease in cold events. Thus we should not yet become overconfident about our views of La Niña's impacts in Cuba. Research on La Niña must improve. Indices and thresholds need to be refined and physical mechanisms of the ENSO cycle better clarified. One should not forget that El Niño was "rediscovered" by some members of the scientific community only after the devastating impact of the 1972–73 event on the Peruvian fishery, even though warm events were known to occur at least as early as the 1600s, because of their impacts on fishing and agricultural activities along the Peruvian coast (Glantz, 1979). Thus, to give La Niña events a real meaning, we need to be more aggressive in finding out what its "fingerprints" are in the economic and social life of our country. If not, it will remain as a marginal product of statistical analysis and of physical hypotheses.

REFERENCES

Cardenas, P. and L. Naranjo, 1996: *Impacto y Modulacion de los efectos ENOS sobre los Elementros Climaticos en Cuba*. Technical report, Instituto de Meteorologia de Cuba.

Glantz, M.H., 1996: *Currents of Change: El Niño's Impact on Climate and Society*. Cambridge, UK: Cambridge University Press.

Glantz, M.H., 1979: The science, politics and economics of the Peruvian anchoveta fishery. *Marine Policy*, 3(3), 201–10.

Gray, W., 1984: Atlantic seasonal hurricane frequency: Part 1. El Niño and the 30 mb quasibiennal oscillation influence. *Monthly Weather Review*, 12, 1649–68.

Hess, J.C., J.B. Elsner, and N.E. La Seur, 1995: Improving seasonal hurricane predictions for the Atlantic Basin. *Weather Forecasting*, 10, 425–32.

Naranjo, L., 1998: Impacts of ENSO in Cuba. In M.H. Glantz (ed.), *A Systems Approach to ENSO: An NCAR Colloquium*. ESIG/NCAR, Boulder, CO, 73–75.

Ortiz, P.L., and A.V. Guevara, 1997: El efecto de un Indice de ENOS en la varibilidad de la serie de *Meningitis Meningococica*. Technical Report. Instituto de Meteorologia de Cuba.

Trenberth, K.E., 1997: The definition of El Niño. *Bulletin of American Meteorological Society*, 72, 2771–7.

The consequences of cold events for Peru

Norma Ordinola

Definition and characteristics

Since 1983, El Niño has been associated in Peru with large ecological and socio-economic disasters, especially in its coastal areas. These impacts have primarily been associated with torrential rains and flooding in northern Peru and devastating droughts in the south, especially across the high Andes. The characteristics of La Niña are generally the opposite to those of El Niño. During La Niña events, the sea surface temperatures are colder than average in the central and eastern equatorial Pacific and the westward-blowing trade winds are stronger than normal. It is fair to state that La Niña, which can also be represented by the positive phase of the Southern Oscillation Index, is not as well understood as El Niño. It is also clear that more attention has been given to the El Niño event and its impacts than to La Niña and its impacts.

Categories of El Niño's intensity

Four categories of El Niño intensity have been defined: strong, moderate, weak, and very weak (Quinn et al., 1978). However, there are considerable differences within each of these categories. For example, while the 1972 and 1976 El Niño events evolved in a similar manner, the events in 1982–83 and 1997–98 developed in different ways. Until recently, El

Niño researchers said that the 1982–83 event was a "mega" Niño, with a probability of a recurrence taking place in a 200–400 year period. However, only 15 years later the world witnessed an even greater El Niño event, which was more intense and had a more severe global impact, in economic terms. The 1997–98 event can be truly called the El Niño of the Century. Presently there are no such categories for labeling the intensities of various La Niña events, and we are not sure what amount of sea surface temperature change is needed in order to produce a strong, moderate, weak, or very weak La Niña.

Predictability

Researchers thought that the Pacific atmosphere-ocean coupling events (El Niño and La Niña) were predictable and that the large general circulation models could predict the occurrences of these phenomena with reasonable accuracy. However, models have some degree of reliability in their projections for only up to three months. The Southern Oscillation is aperiodic and, thus, it is unpredictable. There are also some uncertainties about the intensity to which they might develop, as well as about the nature and magnitude of the impacts associated with these phenomena.

Societal impacts

Societal impacts of both El Niño and La Niña are important. The ripple effect of these events cuts across several geographic and economic boundaries and affects virtually all socio-economic aspects of human activities in Peru. There are both winners and losers as a result of those impacts. Some of the sectors that are most affected are noted in the following paragraphs:

Agricultural impacts

La Niña effects are more pronounced during the Southern Hemisphere winter months (July and August), when one can observe lower than average temperatures. For example, with a few years of low temperatures (around 12 °C), rice crops were adversely affected in 1999. The second harvest yielded only 4000 kg/hectare, while a normal harvest used to be around 7,000 kg/hectare. However, cotton production experiences an increase of 45 percent as a result of cold events. This also took place in 1999 (Figure 3-22).

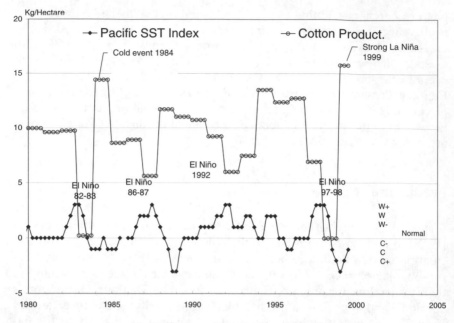

Fig. 3-22 Peruvian cotton production.

Fisheries impacts

Higher yields and, thus, greater exports during cold events have had a positive influence on Peru's fishing industry. For example, in 1996 the anchoveta and sardine catches increased, with a corresponding increase in exports. Warm ocean temperatures are known to be generally associated with a reduction of the catches of some species such as anchoveta. However, when a La Niña event occurs after an El Niño, its positive impact appears to be smaller as was the case during the last La Niña of 1998–2000. This is because La Niña represents different conditions, affecting fish population distribution, ocean composition, and fisheries resource management.

Health impacts

La Niña greatly impacts the health sector as witnessed by an increase in the number of bronchial diseases and respiratory illnesses that occur, especially in locations where the humidity is higher. An example occurred in Piura, in northern Peru, where 200 children developed respiratory illnesses because there was an unusual decrease in temperature in November 2000 (5 °C below average). Health problems also aggravated the already severe poverty and pollution problems, thereby increasing the total sum of human suffering.

Climate impacts

La Niña is usually associated with severe drought conditions in coastal areas as well as with decreased temperatures associated with negative anomalies of sea surface temperature. In fact, an absence of rainfall over the coastal regions of northern Peru was expected during the last strong La Niña in 1998–2000. Suddenly, however, the Piura River discharged a huge volume of water, as had been the case during the 1982–83 El Niño. This was the consequence of intense convective activity in the Amazon region where precipitation increased during the last La Niña event. Only during normal years is there a reasonable amount of rain in the mountainous areas of Peru, as a result of the easterly trade wind flow over the Andes. But there is no significant rainfall along the coastal regions, which is unlike the situation that takes place during El Niño years. Besides, La Niña does not prevent the occurrence of changes in equatorial surface currents. Usually, the sea surface temperatures in the northern part of Peru become a few degrees higher than normal during the summer season, even when it is considered a La Niña year.

Why is there interest in La Niña?

El Niño events and their devastating effects across the globe have received considerable media, scientific, and public attention. However, it is only recently that the rest of the ENSO cycle, which includes La Niña and normal, has come to be examined and analyzed more closely for their characteristics and their consequences. The two extremes (warm and cold) of the ENSO cycle should be studied in a more comprehensive and integrated manner. A piecemeal approach to the ENSO cycle with an over-focus on only one extreme (e.g., El Niño) should be avoided. As for La Niña, it has produced several surprises around the world.

Improved coordination with global climate change studies

In light of the major concerns about global climate change, it will be useful to examine La Niña and El Niño impacts in the broader context of climate change. An integrated and coordinated approach should be successful in bringing together a wealth of knowledge and expertise. Researchers and funding agencies should encourage and support efforts to better understand both extreme changes in sea surface temperature changes in the tropical Pacific and their impacts on society, economy, and environment.

REFERENCES

Arntz, W.E. and Fahrbach, E., 1991: El Niño; *Experimento Climático de la Naturaleza*. Fondo de Cultura Económica, México.

Laws, E.A., 1997: *El Niño and the Peruvian Anchovy Fisheries*. Herndon, VA: University Science Books.

Philander, S.G., 1990: *El Niño, La Niña and the Southern Oscillation*. San Diego, CA: Academic Press.

Quinn, W.H., D.O. Zopf, K.S. Short, and R.T. Kuo Yang, 1978: Historical trends and statistics of Southern Oscillation, El Niño, and Indonesian droughts. *Fisheries Bulletin*, 76, 663–78.

Africa

Kenya and ENSO: An observation and La Niña prediction

Peter E.O. Usher

Awareness of El Niño in Kenya has become virtually universal, as a result of the 1997–98 El Niño. In 1998, eastern Africa suffered unprecedented amounts of rainfall resulting in the spread of infectious disease,
food shortages, death, and infrastructure damage totaling at least in the
hundreds of millions of dollars. Many of these impacts have been blamed
on El Niño, rightly or wrongly. The Kenyan media had daily reports of
adverse events attributed to El Niño, which continued into June 1998,
even though the climate had settled into the normal cold dry season. For
example, on 22 June 1998, the *Daily Nation* reported about a plague of
rats and attributed it to conditions brought about by El Niño. The variety
of rat (*Ratus ratus*) was said to be as large as cats and were blamed
for devastating the maize fields. Separately, it reported the appearance
of a hippopotamus in a Nairobi city stream that had swollen to river
size, following months of torrential rain. In delivering his annual budget
statement in parliament, Kenya's Finance Minister blamed El Niño for
the nation's financial plight. This was an interesting departure from
the government-identified perennial culprits: corruption and government
mismanagement. Important questions result from these attributions –
was the unusual and unanticipated rainfall entirely El Niño-related and
would the La Niña that had been forecast to follow have an influence on
Kenya's weather?

Table 3-7 Seasonality in Kenya

Season	Time period
Hot dry season	Mid-December to mid-March
Long rainy season	Mid-March to mid-June
Cool dry season	Mid-June to mid-October
Short rainy season	Mid-October to mid-December

Kenya's climatology

Kenya has four distinct seasons (see Table 3-7).

As Kenya straddles the equator, the rainy seasons reflect the biannual passage of the Intertropical Convergence Zone (ITCZ) across the country – moving southward early in the year and northward in the second half of the year.

Kenya's dry seasons coincide with the Indian Ocean monsoon flows: warm during the Northern Hemisphere northeast monsoon and cool during the Southern Hemisphere southeast monsoon. This reflects the different origins of the air masses, their fetch over ocean or desert, whether the air is descending or lifted, accelerating or decelerating.

No wet season is entirely wet, and no dry season is entirely dry. Significant rainfall can occur during any month and rainy seasons may fail almost totally in some locations. Their onset and cessation may be early or late, sudden or gradual. The reasons for this large variability from year to year occurs because of the many influences on Kenya's weather – extra-tropical as well as local; oceanic and continental; the symmetry or otherwise of meteorological systems north and south of the equator; the vertical profile of the dominant air masses; and the modifying regional influences of mountains and lakes, particularly Lake Victoria.

Kenya and El Niño

The initial awareness of El Niño in Kenya arose from the experiences of the 1982–83 event. That event was characterized by prolonged dry spells and drought conditions in the West African Sahel and in southern Africa. The plight of those in the Horn of Africa and the farmlands of Zambia and Zimbabwe became familiar to Kenyans by way of the media. Kenya was locally dry. The central highlands had average total rainfall. However, the long rains of 1983 finished a month early and the short rains did not arrive until December, thereby significantly disrupting planting and harvesting and reducing crop yields. December rainfall brought to a close seven months of relatively dry weather, but there was not enough rain to compensate for a further nine months of drought that began in January 1984, which included a complete failure of the March to June long rains. The 1985 short rains were wet but, overall, 1985 was a very dry year (e.g., Cohen and Lewis, 1987).

An examination of all El Niño periods from 1945 to the present indicates neither noticeable drought nor prolonged rainfall except for the recent 1997–98 event, which coincided with a uniquely wet period in the country. Effectively, it had been continuously wet from early October 1997 until mid-June 1998. There was no hot dry season. In fact, record rainfall occurred in many provinces during January and February 1998, with Nairobi experiencing the heaviest rainfall for any month since records began. The one exception occurred in November 1961 which, at 23.05 inches of rainfall, was wetter by 1 inch. However, November is climatically a wet month within the short rainy season with an average rainfall of 6.75 inches; February falls within the hot dry season and has an average rainfall of only 2.5 inches. Although February 1998 rainfall was uniquely high for the season, wet Januarys and Februarys are not unknown. In 1993, for instance, there was 9 inches and 6 inches in January and February, respectively, and the year is also notable for a weather-induced railway disaster which cost more than one hundred lives. There have been only seven wet Februarys and four wet Januarys since 1945. Only in 1958 did a wet February coincide with an El Niño year and only in 1961–62 has there been more than three consecutive months of heavy rain (October 1961–January 1962), again not an El Niño period.

Kenya and La Niña

If no clear relationship can be found between El Niño and climatic variability in Kenya, almost the same can be said of La Niña. Of the seven cold phases since 1945, there have been dry and wet years coincident with La Niña although mostly they have been dry, noticeably so in 1949–50; 1954–56 and in 1984. However, none of these years was as dry as 1976, a non-La Niña year (in fact it was an El Niño year), while the cold episode from 1988 to 1990 was a very wet three-year period. The wettest years on record – 1961, 1963, and 1977 – were neither El Niño nor La Niña years. The driest years recorded are: 1984, 1969, 1954, 1953, and 1949, some of which were years during which La Niña occurred.

Public perception of ENSO

Before the beginning of 1998, few people in Kenya were familiar with, let alone interested in, El Niño. Farmers were the most sensitive to the issue and it is my observation that those acquainted with the 1982–83 and 1987 events generally considered the overall effect to be one of dryness, particularly in the eastern and central highlands and the Rift Valley. The coastal and lake regions were on the other hand wetter than normal. There is considerable spatial and temporal variability in the country's weather at any given period and the climate statistics do not always reveal the vagaries in precipitation distribution.

Individual perception is based on personal experience, which might not be representative of a broader community or region. Nevertheless, as an example, a major Kenyan tea company manager decided against replanting tea crops lost to drought prior to the 1997–98 El Niño, based on an entirely incorrect expectation that the period would mimic the 1982–83 and 1987 situations on the tea estate and would continue to be dry.

The Kenya Meteorological Department's long-range predictions suggested that the early 1997 short rains would continue into the dry season. Although accurate qualitatively, the forecast did not anticipate the amount of rain that fell, nor did the forecast prepare people for the destruction and economic devastation brought about by the rain. Such devastation included the prolonged closure of the only coast-to-interior highway (the Nairobi-Mombasa road), vital for the transit of virtually all goods to and from Central Africa as well as to Kenya's interior. Rail services were also suspended because of landslides, and neither road nor rail service regained their pre-rain condition quickly, which in any case was not particularly good. Makeshift bridges became temporary structures, and metal highways became cratered. Muddy tracks were littered with broken and overturned vehicles.

Initially, the official forecast did not anticipate the continuation of significant rain throughout the April–May 1998 long rains period. In fact, a drought had been predicted, whereas May in particular turned out to be significantly wet and comparable to 1986, 1981, 1980, 1967, and 1958. These were years with above-average May rainfall. Later, forecasts were updated and accurately predicted the arrival of the cool dry weather appropriate to the season.

The imminent La Niña

It is likely that the next El Niño will be anticipated in Kenya with the expectation that it will be comparable with the 1997–98 event, and, therefore, devastatingly wet. However, a review of all El Niño events in the last 60 years suggests that the next El Niño (in the early years of the first decade of the twenty-first century) could as likely be dry as wet, and with only a minimal departure from average conditions.

Based on an examination of all cold events over 60 years, it was probable that the 1998–99 La Niña would be dry, although this could not be predicted with absolute certainty. A clue to what might occur could be found in the prevailing situation in southern Africa. The May 1998 rainfall had been 25 percent below the seasonal average in a broad band from Namibia in the west to Mozambique in the east. This suggested that the Intertropical Convergence Zone (ITCZ) was adopting a more benign character on its northward passage than it had possessed on its earlier southward progression from Central Africa. While the character of the ITCZ can change during its passage north or south, observations suggested that it would be more likely to mimic its profile set at its tropical limits. Hindsight suggests it might have been appropriate to acknowledge the highly active nature of the ITCZ in the Horn of Africa in mid-1997, and to have at least considered the possibility of an above average and even exceptional rainy season in Kenya.

As alluded to earlier, there are many factors which contribute to Kenya's weather patterns. ENSO extreme warm events are among these, but under the majority of situations are unlikely to be the dominant factor. El Niño may or may not have caused the remarkable weather in the first part of 1998, although given the abundance of El Niño-induced weather globally, it would take a brave person to deny the connection. Yet, heavy rainfall has occurred in non-El Niño years, particularly in 1961–62. The year most similar to early 1998 has been 1993, another non-El Niño year. Rather than concentrating on identifying the link between El Niño or La Niña and weather anomalies in Kenya, researchers should also consider the meteorological similarities that might exist during years with comparable weather patterns, irrespective of the existence of ENSO warm or cold events.

Conventional wisdom suggested that 1999 would be dry, as a major La Niña had been forecast for the 1998–99 period. As noted earlier, La Niña events have been coincident with drier periods in Kenya. Public expectation was for weather of an opposite kind to the wet 1998 El Niño. Because this did come to pass, it likely established or reinforced the public perception that forecasting ENSO has improved and that the next El Niño was likely to bring heavy rains to Kenya. Policy decisions, if based on this perception alone, could well prove to be not only inappropriate but expensive as well.

In the event, the prediction of a generally dry 1999 turned out to be predominantly correct, particularly in the north and east of the country. Ethiopia and Somalia suffered extensive and severe drought throughout 1999, and this condition extended southward to the equator. There were many cases of starvation and even death among northern Kenya's Turkana tribe, belatedly acknowledged and responded to by the government following an outcry in Kenya's media. Turkana's normally nomadic way of life usually mitigates the effects of drought conditions. It is probable that the tribe's usual mobility has been restricted by frequent ethnic clashes with other tribes, both within Kenya and with raiders from outside its borders, necessitating clustering rather than dry season disbursement of tribes and their livestock.

Contrary to predictions, the rainy seasons in southwest Kenya and the central highlands were wet. Harvests were adequate, although the transfer of grain to neighboring drought-impacted areas mostly did not occur in a systematic way, with farmers choosing to sell produce for higher prices in the capital and other urban areas, rather than to the needy population in the drought-afflicted north.

The 1999 "long rains" were near normal in the central highlands and the following "short rains" exceptionally heavy, with Nairobi having its fourth wettest November in 60 years. Despite this, the annual average rainfall was low, as the inter-wet-season rain, which accounts for one-third of the annual average rainfall, was generally nonexistent. Reservoir and lake levels fell steadily from their 1998 highs. For example, Lake Naivasha's level in the southern Rift Valley was 2 meters lower by the end of 1999 than it had been at the beginning of the year. Even so, it was still significantly higher (1/2 to 1 meter) than its level in the years immediately prior to the 1998 floods. The national power company introduced power rationing in mid-1999 (which was still in force as of April 2000), restricting electricity supply to six hours per day on three days each week to all consumers, both domestic and commercial. The economic implications of this are almost catastrophic, given the simultaneous withdrawal or restriction of international aid by western donors and the Bretton Woods' Institution, as a result of concerns about governance and corruption.

Kenya's primary foreign exchange earners were also under performing. Concerns over crime and violence had decimated the tourist industry and a combination of weather-related events and marketing mismanagement severely handicapped the coffee and tea industries. For example, tea yields, severely reduced by drought in mid-1999, were further reduced by unprecedented frost in early 2000; up to 15 percent loss occurred of mature and developing tea bushes on some highland tea estates in Kericho and Nandi.

Lack of local resources and restricted international aid have meant that much of the "El Niño flood damage" to infrastructure had not been fully repaired by the turn of the millennium, further adding to the economic woes of the nation. The unkind weather, coincident with the La Niña event, had exacerbated an already serious socio-economic and political situation.

The increased activity of the ITCZ, early in 2000, after its southwards passage across the equator, was maintained throughout southern Africa. Indeed, supported by an unusual number of severe tropical cyclones making landfall in Madagascar and Mozambique, the rainy season in those countries may have been uniquely heavy. Widespread flooding resulted in hundreds of deaths and economic devastation and necessitated a concerted international disaster-response effort to control.

The ITCZ moves northward during March and April. Kenya would expect its long rainy season to begin at the end of March 2000. The forecasts issued by the Kenyan Meteorological Department (KMD) and the Drought Monitoring Centre in Nairobi anticipated continued rainfall deficit for the country as a whole with the drought-afflicted areas of the east and the north enjoying normal or above normal rainfall.

New investigations of teleconnections are essential, particularly for those areas where the ENSO warm and cold event signals are more capricious than for those areas where the regional climate response is both relatively automatic and predictable. Public information on ENSO's extremes also needs to be carefully tailored to take account of and make explicit the many uncertainties surrounding the event and of the rather limited ability to make accurate predictions.

REFERENCE

Cohen, J.M. and D.B. Lewis, 1987: Role of government in combating food shortages: lessons from Kenya 1984–85. In M.H. Glantz (ed.), *Drought and Hunger in Africa*. Cambridge, UK: Cambridge University Press, 269–296.

The impact of cold events on Ethiopia

Tsegay Wolde-Georgis

Ethiopian agriculture is based on the activities of small-scale peasant farmers. The production of the country's food and agricultural exports depends on them. Ethiopia in turn depends on rain-fed agriculture with limited use of irrigation. There are three seasons in Ethiopia, classified by the amount of rainfall: the main rainy season, the *kiremt* (June–September); the dry season, *bega* (October–January); and the small rainy season, *belg* (February–May). Climate-related hazards such as droughts or floods significantly impact agricultural output, which makes Ethiopia vulnerable to climate variability from one year to the next. It also creates an urgent need for the integration of climate variables into agricultural planning at various levels of society. Reliable climate information can also help policy makers to provide users (e.g., farmers, pastoralists) with timely early warning about droughts, floods, fires and other climate-related disasters in order to reduce their impacts.

One of the key goals of the Ethiopian government at the beginning of the twenty-first century is food self-sufficiency. The government provides inputs such as fertilizer, seed, and extension services to farmers who are encouraged to participate in water and soil conservation activities. These policies have already produced impressive results in food production, such as the record harvests of 1996–97 and 1998–99, which brought food security at the national level for the first time in many years. Favorable weather conditions contributed to the record harvest during these two seasons. Adequate and credible dissemination of weather information to

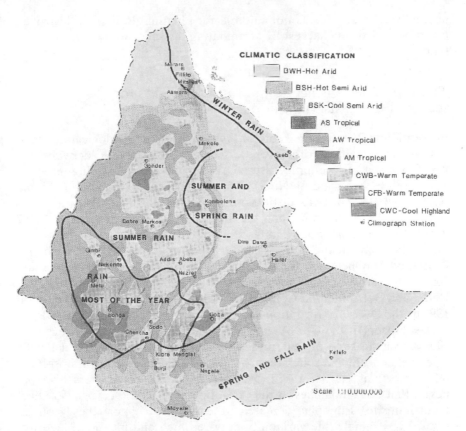

Fig. 3-23 Climatic zones of Ethiopia (from Ethiopian Mapping Authority, 1988).

farmers by the government can strengthen the positive effects of agricultural policies on food security.

Various Ethiopian governments have tried to use climate information for food security. Climate planning in Ethiopia has depended on statistical averages of rainfall and temperature information from a limited number of meteorological stations in the country. However, conclusions based on these temperature and rainfall averages from the various stations can be deceiving. The stations are limited in number (about 600 in March 2000) and Ethiopia has diverse microclimates (NMSA, 1998).

Agricultural drought could be caused by variations in rainfall, temporally or spatially. The effectiveness of rainfall, however, depends on the length of the rainy season, rainfall distribution, timing, and intensity. For example, Ethiopia's unusually heavy rainfall in November 1997 was of little use for food production activities, because its timing was during the harvest season and it was not suitable for planting. Rather, it destroyed much of the existing harvest by germinating the seeds well before harvest time.

The use of sea surface temperatures (SSTs) for seasonal forecasts in Ethiopia

In addition to its traditional role of inventorying meteorological information, the National Meteorological Services Agency (NMSA) actively uses worldwide sea surface temperature (SST) conditions (including information about El Niño and La Niña) to forecast seasonal rainfall in Ethiopia (Glantz, 2001). This was institutionalized after the mid-1980s, as a result of the 1983–84 famine and the 1982–83 El Niño. It was in the 1987 drought that El Niño/Southern Oscillation (ENSO) information was used for the first time by the NMSA as an early warning for possible problems with food production activities and for famine avoidance (Kassahun, 2000). The early warning was distributed to users such as the government's Relief and Rehabilitation Commission (RRC). The RRC was renamed in 1995 as the Disaster Prevention and Preparedness Commission, or DPPC.

Researchers at the NMSA and policy makers in Ethiopia believe that changes in the state of sea surface temperatures in the tropical Pacific Ocean affect Ethiopian climate through statistically identified atmospheric teleconnections (Haile, 1988). Moreover, NMSA researchers also believed that SST anomalies in the Atlantic and Indian Oceans also affect (contribute to) Ethiopian climate. Other studies support these conclusions. For example, the variability between high and low flood levels of the Nile River, whose major sources are in the highlands of Ethiopia, has been related to the ENSO cycle (e.g., Quinn, 1992; Flohn and Fleer, 1975). Ethiopia has since used El Niño information as an early warning for the 1987, 1992, and 1997 droughts. It has also used La Niña information to forecast *kiremt* floods for 1996, 1998, and 1999. Based on the information provided by NMSA, the DPPC was able to disseminate early warnings to users.

Recent La Niña impacts on Ethiopia

Recently, Ethiopian policy makers have begun to incorporate La Niña information into their early warning system as well. Mr. Teshome of the DPPC's Early Warning Department (EWD) confirmed that they used La Niña information to prepare for various climate-related hazards, such as droughts and floods, since 1996 (Teshome, 2000).

Until recently, the focus of policy makers on cold events and their impacts was much less than their focus on El Niño. The main (and chronic) cause of food insecurity and disasters in Ethiopia was drought, and so no one was ready to take a forecast of heavy rainfall very seriously. Despite this, the DPPC and NMSA began to use La Niña-based early warning information in 1996.

Analysis of past climate conditions in Ethiopia reveals that during La Niña events, the following weather patterns occur in Ethiopia (DPPC, 2000). These include: (1) late onset, below normal amount and inadequate rainfall distribution in the *belg* crop growing areas; (2) poor and erratic rainfall in the pastoral areas of the south and southeastern Ethiopia (the Ogaden and Borena pastoralists); and (3) above-normal rainfall during the *kiremt* season. Generally, La Niña events disrupt the distribution of the small rainfall system during October–January and February–May. One of the reasons for the poor performance of rains in September–December may be the withdrawal of the rain-bearing Inter-Tropical Convergence Zone (ITCZ), which brings moist air to Ethiopia from the pastoral areas of Ethiopia (DPPC, 1999). The pastoralists depend on these rainfall systems for water and grazing for their herds. These areas are normally dry, during the main *kiremt* rainfall in Ethiopia.

The small rains are important in the *belg*-producing areas that are located at the eastern escarpment of the Ethiopian plateau west of the Ethiopian Rift Valley. During the *belg* season, farmers plant short-maturing crops such as wheat, barley and teff (a local staple grain) that contribute between 5–10 percent of the national annual output. At the local level, however, production may be as high as 40–50 percent of a household's annual food production. The *belg* precipitation is also used by farmers to prepare the land for *kiremt* planting. *Belg* rains replenish water and pasture in the dry areas.

The recent cold events of 1996 and 1998–2000 affected the pastoral regions and *belg*-producing regions of Ethiopia in various ways. The Ethiopian early warning system had issued meteorological forecasts about the climate hazards caused by the cold events.

According to the DPPC report, the October–November 1996 short rains over the Somali region, Borena zone and Hamer Bena Wereda of SNNP regions were much below normal, erratic and unevenly distributed. The late arrival of the small rains in February–April resulted in critical and widespread water and pasture shortages. Consequently, human lives were at risk and livestock conditions deteriorated rapidly. Rivers, ponds, wells and other watering points dried up. All these contributed to high livestock losses (DPPC, 5/1997: 3).

The 1996 *kiremt* rainfall began in June 1996 as normal and was particularly heavy in the northern highlands of Tigray and eastern Amhara. Some torrential rains also caused flooding and property damage. Most parts of Ethiopia received 25 to 30 days of rainfall, which is a good temporal distribution, and some of the stations had recorded heavy rainfall in one day (*ibid.*). This excellent *kiremt* rainfall contributed to the record harvest of 1996–97.

According to Mr. Bokretsion Kassahun of NMSA, the pastoral and *belg* crop areas of Ethiopia had been under a continuous dry spell since the appearance of the prolonged cold event in 1998 (Kassahun, 2000). The cold event that began in July 1998 persisted for the last three years (Monastersky, 1999: 278). In 1998, the Early Warning Report of the DPPC stated that: "The onset of the *kiremt* rain was late in the northeast and east and timely in the south and western parts of the country" (DPPC, 6/7, 1998). However, it issued a forecast for the month of August 1998 and concluded that the *kiremt* rainfall regions will continue to receive their normal rainfall (DPPC, 5/7, 1998). This was good news for farmers, because of the dry June in 1998 and the late start of the rainy season. In August 1998, there was an improvement in the amount and distribution of rainfall in most parts of Ethiopia. Major flooding and landslides occurred in some parts of the country (DPPC, 8/1998: 1). However, the October rainfall was below normal in most parts of the pastoral areas of the Afar and Somali Regions (NMSA, 1998: 1; FEWS, 11/1998).

With the exception of some lowland areas, almost all parts of the country that receive *kiremt* (summer) rains had in general normal to above-normal rainfall during August 1998 (DPPC, 8/1998: 1). The 1998 *kiremt*, "which provided adequate rainfall over a sufficiently long growing season and successful agricultural extension programs that provided farmers with fertilizers, seeds, and recommendations of timing of weeds and other production practices," produced Ethiopia's second record harvest in 1998 (FEWS, 11/1998). FEWS stated that the 1999 food-aid needs could be met through local purchases rather than imports (*ibid.*). Some donors even stated that domestic consumption alone would not absorb all the surplus production and recommended the export to food-short neighboring countries such as Kenya and Somalia (FEWS, 12/1998). Despite the surplus production, the lowland and pastoral areas of the Harerghe, Borena, and Somali regions were affected by a food deficit because of localized drought. Drought-related cattle deaths became widespread in 1999–2000, as reported by the DPPC in 1998–99 (FEWS, 1/1999).

An initial early warning report for the 1999 *belg* season stated that the bad performance of the *belg* crops was expected to worsen the food situation in the country (FEWS, 3/1999). It also mentioned that the drought situation in the pastoral areas of the country was reaching disaster proportions. In April–May 1999, the NMSA issued a long-range forecast indicating that the 1999 *kiremt* season would be under the influence of

La Niña. The forecast stated that the La Niña event would continue to prevail until September, and that the *kiremt* rainfall areas in the country would receive normal rainfall until September 1999 (DPPC, 4/5, 1999). It also gave a forecast stating that parts of Tigray and Amhara regions were "expected to receive above-normal rainfall." The early warning advised that there would be heavy rainfall with hailstorms and possible floods in some parts of the country. Its forecast also stated that the pastoral areas, which receive their rains in October–December would receive their rainfall earlier than normal starting in mid-September (DPPC, 4/5, 2000: 10). Reports in the Somali region of Ethiopia (including Jigiga, Dagahabur, and Warder zones) were experiencing scarcity of water and pasture. In fact, the Somali region was planning to transport water to the affected areas of the Degahabur and Warder zones (DPPC, 4/5, 1999: 4).

The 1999 *kiremt* rainfall was good in most parts of the country and reflected the NMSA's seasonal forecast. In its July–August 1999 issue, the Early Warning System's *Monthly Report* stated that "the rainfall covered nearly all the *kiremt*-rainfall-benefiting areas of the country." It also stated that, "During the first week of August, rainfall conditions further improved significantly and covered nearly all the *kiremt* rainfall-benefiting areas of the country." On the other hand, there was below normal rainfall in the pastoral areas of eastern Ethiopia, such as the northern Somali region and the eastern escarpment of Ethiopia (DPPC, 4/5, 1999).

Cereal production in 1999 was affected by the failure of the *belg* rains. The poor *belg* rainfall did not favor the planting of the long-maturing crops such as maize, sorghum, and millet. Long-cycle crops that were planted in April 1999 dried up or were wilting. The *meher* rains started in the last week of June and were favorable for short-maturing crops; the long-maturing crops failed. Farmers attempted to replant the failed crops with short-maturing crops. There were shortages of seeds, and the government intervened and provided 7,860 MT of seeds at a cost of 29,138,920 *Birr* in response to the shortages (DPPC, 4/5, 1999: 3). There was surplus food production in the 1999–2000 agricultural season.

The weather conditions in Ethiopia between February and March 2000 were dry. The February 2000 Early Warning System's *Monthly Report* stated that "the current *belg* season would continue to remain under the influence of La Niña, indicating that the La Niña will persist at least through the first half of 2000" (DPPC, 2/2000: 1). The *belg* rains failed and did not start until the beginning of May 2000.

Mr. Teshome, who heads the DPPC's Early Warning System, noted that the La Niña episode that has prevailed in the Pacific since July 1998 has been doing an abundance of damage in Ethiopia (Teshome, 2000). He stated that La Niña disrupts the rainfall pattern of the *belg*-producing and pastoral areas of the country. These areas include the escarpments west of the Ethiopian Rift Valley, and the Somali (Ogaden) and Borena regions. The impact of the recent La Niña had been felt in these Ethiopian areas for three years. The areas that were hardest hit by La Niña were those that normally get their rainfall during the small rains of October–December and February–April (Kassahun, 2000).

Most of the *belg*-producing regions of northern Shoa, north and south Wello and southern Tigray produce a major portion of their crops during the *belg* season (Kassahun, 2000; Teshome, 2000). In the normally *kiremt*-dependent zones of Ethiopia, long-maturing crops such as maize, sorghum, and other species of barley and teff, take advantage of the *belg* rains to complete their maturation at the end of the *kiremt* season. Therefore, the disruption of precipitation, due to La Niña, affects the Ethiopian economy.

The disruption of the small rains pushes the planting of the long-maturing crops. These crops are planted in April and May and harvested after the end of the *kiremt* rainy season. But the people that were the worst affected by the recent La Niña had been the pastoralists of eastern and southern Ethiopia. The recent looming disaster in the Somali region of Ethiopia was clearly caused by the persistent three-year drought. The three-year drought dried out water wells and changed the land surface into dust. The aridity first killed cattle, and later began to kill sheep, goats, and then even camels. A shortage of food and water for people and animals has become one of the acute problems in the area. The 1999–2000 disaster was averted by heavy international intervention.

Because of the normally erratic nature of the *belg* rains and the influence of other climatic factors, we have to be careful in concluding that all La Niña years disrupt the *belg* rains or that all El Niño years lead to above-normal rainfall during the *belg* season. As the NMSA has noted, other atmospheric and oceanic interactions such as those in the Indian and Atlantic Oceans and associated atmospheric pressure affect the timing and distribution of precipitation in Ethiopia.

Therefore, in general, the major impacts of La Niña on Ethiopia are a dry *belg* (small rains) season and an above-normal *kiremt* rainfall. As noted earlier, there was an absence of the small rains during the 1996, 1999, and 2000 cold events. Cold events also caused heavy rainfall and flooding during the 1996, 1998, and 1999 *kiremt* (long rain) seasons and, as forecast by the NMSA, there were crop damages in many parts of Ethiopia, because of this excessive rainfall (DPPC, 4/5, 1999).

Conclusion

The relationship between Ethiopian weather conditions and the ENSO warm event/cold event cycle is not yet conclusive, because of the limitations in data and the diversity of the Ethiopian microclimates. However, there is a statistical association between Ethiopian droughts and El Niño (Wolde-Georgis, 1997; WMO, 1984). Being in an El Niño year, the 1997 *kiremt*, for example, was unusually dry in Ethiopia, and the reservoirs around Addis Ababa were unusually low in 1997 (NMSA, 1997). November is typically the driest month in the central highlands of Ethiopia, but in 1997 it was unusually wet. This anomaly may have been related to or influenced by the major El Niño of 1997, as well as the unusually warm SST anomaly in the western Indian Ocean.

According to the NMSA, cold events cause less than average rainfall in the *belg* season and heavy rainfall during the *kiremt* season. Reports from Ethiopian media during the 1996, 1998, and 1999 *kiremts* noted exceptional floods in many parts of the country. Many rivers, including the important Awash River, have flooded their banks, which led to the evacuation of many settlements in the Ethiopian Rift Valley. Thus, the 1996 and late 1998 La Niña events were accompanied by unusually heavy rainfall during the Ethiopian *kiremt* season. The early warning reports of the DPPC also showed that the 1998, 1999, and 2000 *belg* rainfalls were less than normal in the *belg*-producing agricultural and pastoral areas (DPPC, 1998, 1999, 2000).

The impacts of the cold events on the pastoral areas of Ethiopia had been visible since the fall of 1998. The DPPC and the FEWS had provided their early warnings about the impact of this localized drought in 1998, 1999, and 2000 (FEWS, 12/1998; FEWS, 1/1999; DPPC, 2000). The DPPC was moving food to the Somali region, but there were unsolved problems of water and fodder shortages (FEWS, 2/1998).

Policy and other decision makers need to take advantage of the forecasts of cold events and incorporate them into their decision-making activities. For example, farmers could be encouraged to plant seeds that require heavy rainfall and long-maturing periods. Water conservation measures, such as micro-dams, could harvest the heavy *kiremt* rainfall during La Niña years for future use. In addition, flooding and erosion prevention measures could also be put in place ahead of the rainy season, as a mitigating strategy.

Unique local conditions must be taken into consideration in any policy recommendation, so that future early warnings could be both reliable and effective. Ethiopia is characterized by diverse microclimates because of its tropical mountains and lowland deserts. Implementation of responses to a reliable but generalized forecast that covers the whole country would be difficult because of the sharp local differences in microclimates. However, Pacific and other sea surface temperature information could still be used for policies at the national level. These policies include agricultural strategies, such as increasing food security reserves, changing exports and import policies, and consideration of foreign assistance. The people that are most impacted by drought are peasant producers and herders who depend directly on rainfall for their livelihood.

Ethiopia has an adequate drought early warning system that includes the use of ENSO information to forecast drought. However, the capacity of the country to respond to drought is not yet strong. This can be strengthened by introducing controlled irrigation in the cereal-producing areas and through the creation of dams that could be used for irrigation during water shortages. Ethiopia has the largest cattle population in Africa. But the cycle of drought sharply depletes the population of animals. The government has to create a mechanism to educate the pastoralists on when to sell their animals and also on how to participate in modern ranching systems. An adequate relationship with the market could also help the nomads to sell their animals at fair prices, during drought and non-drought years.

Research and policies should focus on the impact on local weather of all ENSO-related SST considerations, including La Niña. Until now, policy makers and researchers considered El Niño as abnormal and all the other parts of the ENSO cycle (i.e., average SST conditions and La Niña) as normal. As a result, cold events have received very little research attention. This attitude is slowly changing with the spread of ENSO information and education and with expanded national and global media coverage. Ethiopia is vulnerable to climate variability from year to year, and its decision makers at all levels of society need to improve their understanding of SST anomalies and their potential impacts on the country, in order to attain their desired, high-priority goal of improved food security.

REFERENCES

Addis Zemen (Ethiopian Amharic daily): Heavy floods expected in some parts of Ethiopia (Hamle 3, 1998, Ethiopian Calendar), 10 July 1998.

DPPC (Early Warning Department), May 1997: *Progress of the 1997 Belg season.* Ethiopia: Addis Ababa.

DPPC (Early Warning Department), June 1997: *Early Warning System Monthly Report.* Ethiopia: Addis Ababa.

DPPC (Early Warning Department), May 1997: *Early Warning System Report, Food Situation and Assistance Requirements in 1997.* Ethiopia: Addis Ababa.

DPPC (Early Warning Department), June/July 1998: *Ethiopian Early Warning System Monthly Report.* Ethiopia: Addis Ababa.

DPPC (Early Warning Department), August 1998: *Ethiopian Early Warning System Monthly Report.* Ethiopia: Addis Ababa.

DPPC (Early Warning Department), August 1998: *Early Warning Systems Report: 1998 Belg Production and Food Prospects.* Ethiopia: Addis Ababa.

DPPC (Early Warning Department), February/March 1999: *Ethiopian Early Warning System Monthly Report.* Ethiopia: Addis Ababa.

DPPC (Early Warning Department), April/May 1999: *Ethiopian Early Warning System Monthly Report: The Impact of the Belg Rains.* Ethiopia: Addis Ababa.

DPPC (Early Warning Department), April/May 1999: *Ethiopian Early Warning System Monthly Report.* Ethiopia: Addis Ababa.

DPPC (Early Warning Department), February 2000: *Ethiopian Early Warning System Monthly Report.* Ethiopia: Addis Ababa.

Ethiopian Mapping Authority, 1986: *National Atlas of Ethiopia.* Addis Ababa: Ethiopian Mapping Authority, 16.

FEWS, November 1998: *FEWS Bulletin.*

FEWS, 30 December 1998: *FEWS Bulletin.*

FEWS, January 1999: *FEWS Bulletin.*

FEWS, 29 January: *FEWS Bulletin.*

FEWS, 26 February 1999: *FEWS Bulletin.*

FEWS, 31 March 1999: *FEWS Bulletin.*

Flohn, H. and H. Fleer, 1975: Climatic teleconnections with the equatorial Pacific and the role of ocean/atmosphere coupling. *Atmosphere*, 13, 96–109.

Glantz, Michael H., 1996: *Currents of Change.* Cambridge: Cambridge University Press.

Glantz, M.H., 2001: *Currents of Change: Impacts of El Niño and La Niña on Climate and Society.* Cambridge, UK: Cambridge University Press.

Haile, Tesfaye, 1988: Causes and characteristics of drought in Ethiopia. *Ethiopian Journal of Agricultural Sciences*, 10(1–2), 111–20.

Kassahun, B., 2000: *Interview*, 11 May. Ethiopia: Addis Ababa.

Monastersky, R., October 1999: La Niña will whip up U.S. winter weather. *Science News*, 156, 31.

NMSA (National Meteorological Services Agency), 1996: Climate and Agro-climatic Resources of Ethiopia, *Meteorological Research Report*, 1(1). Ethiopia: Addis Ababa.

NMSA, November 1997: *Early Warning Systems Report: The Impact of Abnormal Weather in 1997.* Ethiopia: Addis Ababa.

NMSA, 1998: *Annual Climate Bulletin.* Ethiopia: Addis Ababa.

Quinn, W., 1992: The study of Southern Oscillation-related climate activity from AD 622–1900 incorporating Nile River flood data. In H.F. Diaz and V. Markgraf (eds.), *El Niño, Historical and Paleoclimatic Aspects of the Southern Oscillation.* Cambridge, UK: Cambridge University Press, 119–49.

Teshome, A., 2000: *Interview*, 8 May. Ethiopia: Addis Ababa.

Wilhite, D.A. and M.H. Glantz, 1985: *Planning for Drought: Toward a Reduction of Social Vulnerability.* Boulder, CO: Westview Press.

WMO (World Meteorological Organization), 1984: *The Global Climate System: A Critical Review of the Climate System During 1982–84.* Geneva: WMO.

Wolde-Georgis, Tsegay, 1997: El Niño and Drought Early Warning in Ethiopia. *Internet Journal of for African Studies*, 2, March. wwww.brad.ac.uk/research/ijas/ijasno2/ijasno2.html.

La Niña and its impacts in South Africa during 1998–2000

Mark Majodina

La Niña effects on South African rainfall

The Pacific cold event (La Niña) is associated with variable climate conditions all over the world. In southern Africa, it is generally associated with wet and cold conditions. In a climatological study of South African rainfall for the period 1951–95, it has been shown that near to above-normal rainfall occurred in 80 percent of La Niña years. This, therefore, shows that dry conditions do sometimes occur, even during La Niña events.

La Niña events in South Africa (SA) are also generally associated with below normal temperatures, particularly in the Southern Hemisphere mid-summer months (December–February). Earlier studies of the atmospheric circulation anomalies during La Niña years by Lindesay (1986) showed the effect of the Walker circulation to be strongest during late summer months (January–March). On the spatial distribution of La Niña's influence on SA rainfall, Tyson (1987) found a positive correlation between summer rainfall and the Southern Oscillation Index (SOI), over most parts of South Africa except the South Western Cape. The South Western Cape is a winter rainfall region with a rainfall peak during May–July and is mainly affected by mid-latitude westerly systems. The positive correlation (rainfall and SOI) was particularly strong over a northwest to southeast belt across the central areas of the country (see Figure 3-24).

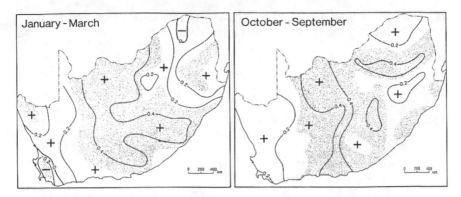

Fig. 3-24 Map showing the correlations between SOI and rainfall in South Africa (after Tyson, 1986).

The rainfall-SOI relationship was, however, found to be relatively weak and accounted for approximately 20 percent of rainfall variance over central South Africa.

Located at the southern tip of the African continent, South Africa is surrounded by a large ocean environment, with three distinct ocean current regimes: the Agulhas Current system along the south and east coasts, the Benguela Current system along the west coast and the subtropical convergence, south of the subcontinent (approximately 38–42°S). When studying the interannual rainfall variability in South Africa, it is necessary to consider the influence of the local sea surface temperatures (SSTs). Lutjeharms and van Ballegooyen (1988) have shown that the warm waters from the Agulhas Current, which flow poleward along the southeast coast of Africa, flowed westward during wet summers, owing to changes in the wind stress field. This coincided with anomalously low SST in the tropical Southeast (SE) Atlantic. Significant correlations were also found between South Africa's summer rainfall and the central equatorial Indian Ocean (CEI) SSTs (Jury and Pathack, 1993). Cadet (1985), in earlier works, showed the lag relationships and possible transmission of ENSO signals between the tropical Pacific and the Indian Ocean SSTs. The SST cooling in the Indian Ocean which is consistent with the Pacific cooling (e.g., the global La Niña event) and the associated atmospheric circulation anomalies are currently thought to be the manner in which ENSO signals are transmitted to southern Africa (Jury and Pathack, 1993). The cooling in the CEI enhances subsidence and east-west overturning and suppresses the uptake of cross-equatorial monsoon flow by tropical cyclones. The moisture, therefore, becomes available for westward advection into southern Africa.

A comparison of the 1998–99 rainfall forecast with the corresponding observed rainfall

The seasonal forecasts of the South African Weather Bureau are based on both statistical (Canonical Correlation Analysis, Optimal Climate Normals, Quadratic Linear Analysis) and dynamical models (Center for Ocean-Land-Atmosphere Studies [COLA], General Circulation Models). The prominent equatorial Pacific cooling featured strongly in the period prior to construction of the 1998–99 rainfall seasonal forecast.

October–December (OND) 1998

The initial forecast for the OND 1998 season was constructed in August 1998 and showed near to below-normal rainfall over the Western Cape and western coastal regions. In the central parts of the country, near normal rainfall was forecasted with a slight bias towards the above-normal category. In the Lowveld region (far northeastern parts of South Africa), near to below-normal rainfall was expected. The South African provinces are shown below in Figure 3-25.

Fig. 3-25 Figure showing the nine provinces of South Africa.

A second forecast was released after the SARCOF (Southern African Regional Climate Outlook Forum) meeting in September 1998. In SARCOF meetings, seasonal forecasters from meteorological offices in the Southern African Development Community (SADC) region and also from leading international meteorological institutes such as UK Met Office, International Research Institute (USA), and Climate Prediction Centers (NOAA) assemble to create consensus forecasts for the SADC region. This updated forecast maintained the near to below-normal rainfall category for the Western Cape Province, western coast and southwestern parts of Namibia. Near-normal rainfall was anticipated over the rest of the country.

The observed rainfall during this period was mainly in the near-normal category. Above-normal rainfall was recorded in isolated areas (Free State Province, southern parts of the North West Province).

January–March (JFM) 1999

The models which were initialized in October 1998 all concurred in response to the prominent equatorial Pacific cooling and called for near to above-normal rainfall over the entire country. After the SARCOF Correction meeting in December 1998, an adjusted JFM forecast for South Africa was constructed which showed near-normal rainfall over the Western Cape Province, above-normal rainfall over a northwest-southeast band in the central parts of the country and near to above-normal rainfall over the eastern half of the country (Gauteng, Mpumalanga, Free State and North West Provinces).

The observed rainfall for JFM 1999 was, however, quite different from the forecast. Most of the country received below-normal rainfall. Near and above-normal categories of rain were received only in some isolated areas such as southern coastal regions, KwaZulu-Natal, and the eastern parts of the Northern Province.

October–December (OND) 1999

Above-normal rainfall was predicted for the western half of the country (Western Cape, North Cape, and the western sector of the Eastern Cape Province). Further east, near-normal rainfall was expected over the Free State, North West, Northern Province, Mpumalanga, and the eastern sector of the Eastern Cape Province.

The observed rainfall revealed the accuracy of the forecast. The entire western half of the country was indeed above-normal, while near-normal rainfall was observed over the eastern half. Some isolated areas in the eastern half of the country (eastern parts of the Northern Province and northern sector of Mpumalanga), however, received above-normal rainfall.

January–March (JFM) 2000

Below-normal rainfall was forecasted for the southwestern Cape, while near- to above-normal rainfall was anticipated over the rest of the country.

The available December 1999–February 2000 (DJF 00) observed rainfall suggests a below-normal rainfall category over the western coastal region, while the rest of the country is experiencing above-normal rainfall. Some isolated areas such as the northwestern parts of KwaZulu-Natal and northern Swaziland, have, however, recorded below-normal rainfall.

In summary, the seasonal forecast for summer rainfall during the 1998–99 and 1999–00 La Niña periods have been relatively accurate. An exception is made for the JFM 99 forecast which was completely misleading. The atmospheric circulation then appeared to have been completely unresponsive to the predominant ENSO signal.

Impacts

Anomalous climatic patterns in the semi-arid SADC region have huge impacts on the regions' food resources, on the agriculturally based economy, and on the general well-being of the population. South Africa in particular (the wealthiest country in the region), however, has a large proportion of its population living in the poor remote rural areas and also in informal settlements.

October 1998

During this month, no extreme weather events were observed. The month was characterized by a succession of swift-moving cold fronts, which seldom penetrated inland of the coastal regions.

November 1998

11/03   Flooding in Welkom (Free State) caused extensive damage to houses.

11/19   More than 80 people were left homeless in Tzaneen (Northern Province), after heavy rains destroyed their houses.

11/20   150 people were evacuated in Evaton (Gauteng), as the Rietspuit River overflowed.

11/24   Approximately 600 people were left homeless, when the Orange River burst its banks at Ritchie (Northern Cape).

December 1998

12/01   Heavy rain over the Bethlehem area in the Free State caused the Jordan River (which flows through the town) to burst its banks and cause floods to the surrounding buildings. The Loch Athlone dam, which is a source to this river, also overflowed and flooded the Bohlokong township, destroying 300 houses.

12/14   Unseasonal heavy rains in the Western Cape (a winter rainfall region) destroyed a bridge along the N2 national road. The national road between Caledon had to be temporarily closed. Strong winds in Cape Town destroyed trees and disrupted telephone communications and power supplies.

12/15   A tornado moved through Hogsback (Eastern Cape), destroying property and killing at least one person. Later, a devastating tornado struck Umtata (also in the Eastern Cape), killing 17 people and injuring over 160 people. Damage to property was enormous. The area was subsequently declared a disaster area. On the same day, a tornado hit Klipriver (Gauteng) destroying houses and a nearby informal settlement of approximately 400 families. Another tornado was reported in Pomeroy (KwaZulu-Natal) which killed one person and caused extensive damage to property.

January 1999

01/11   Heavy storms in the Pinetown area (KwaZulu-Natal) destroyed an electrical substation. Road, rail, and air traffic were also disrupted in the Durban area. Houses were flooded and roads closed as a result of deep water and mudslides.

01/14   Four people were killed by lightning in three separate incidents near Thohoyandou (Northern Province). At Libode (Eastern Cape), seven people were killed when lightning struck the house in which they were sheltering.

01/19   A devastating tornado in the Mount Ayliff area (Eastern Cape) hit six villages destroying and damaging houses. 22 people died in this event, while a further 360 were reported to have been injured. Livestock were also killed and crops destroyed. The area was later declared a disaster area.

01/22   A violent storm caused extensive damage to 20 houses and a church in the Sterkspruit area (Eastern Cape). Houses were later destroyed and electrical wires torn loose in the neighboring Mount Frere area.

February 1999

02/02   Following the heavy rains at the end of January, Tzaneen (Northern Province) was hit by floods and mudslides. South African Air Force helicopters were called in to help, when the neighboring village of Ga-Sekororo was washed away by mud and floodwaters. At least six people died during these floods. In the Pietermaritzburg area (KwaZulu-Natal) several informal settlements were destroyed by heavy rains. Two people were reported to have drowned during this incident. Two women were struck by lightning in Soshanguve (Gauteng), while walking on a footpath.

02/04   Torrential downpours caused considerable damage in the Durban area. Several houses and roads were flooded, rivers and streams overflowed their banks, power supplies and telephone communications were disrupted and hundreds of people were left homeless. Seven people lost their lives during this event and, once more, the informal settlements sustained the heaviest damages.

02/24   Three women were struck by lightning in Plooysberg (Northern Cape), while walking from work.

March 1999

Heavy rains which had been falling in Mozambique since December 1998, have caused flooding over large areas of the Inhambane Province. Many bridges and roads were washed away by the floods which have been described as the worst in 40 years. High temperatures and low (scattered) rainfall generated dry conditions in the maize-growing regions (Free State). The loss to the maize harvest was estimated to be 25 percent.

October 1999

10/21   A tornado caused an estimated damage of 2 million rands when it hit a farm near Greylingstad. Farm buildings were destroyed and livestock were killed during this event. Around Vereeniging (Gauteng), approximately 300 housing units were damaged, when a violent storm hit the area. There were also widespread reported electricity failures.

10/25–26   Extensive damage was caused by floods in Durban and estimated at several million rands. Damage to Virginia airport was caused by 375 mm of rain. Four people were reported to have drowned, while 13,000 homes and businesses were without electricity following the flooding of substations and damage to power lines. Hundreds of residents in informal settlements around Durban were left homeless after this event.

10/27   Three people died and several were injured when a hailstorm hit the Tzaneen and Giyani areas (Northern Province). Telephone and electricity lines were blown over and roads were blocked by uprooted trees.

November 1999

11/06   Five people were injured and another 10 were injured when two houses were struck by lightning at Elliotdale (Eastern Cape).

11/15   About 200 families were left homeless in the Mpophomeni informal settlement, after a severe flood event. At least one person drowned and 35 were treated for injuries and hypothermia.

11/16   Lightning struck a hut in Lothair (KwaZulu-Natal) killing a woman and injuring seven children.

11/19   Three children near Messina (Northern Province) were struck by lightning, while walking home in an open field.

11/23   Ten people drowned near Beit Bridge, after their minibus taxi was swept from a bridge over a flooded river.

December 1999

12/06   In Tzaneen (Northern Province) 77 mm of rain was measured which led to the destruction of a shopping complex, several homes, and a bridge (over the Letaba River). In neighboring Elim, 40 houses were flooded and about 200 people left homeless.

12/06   A heat wave, which dominated the Cape Metropolitan area (Western Cape) and lasting nine consecutive days, resulted in water shortages and an introduction of water usage limits by municipalities. At least four dogs died due to dehydration and others were treated after collapsing.

12/09–28   At least 23 people died in Durban because of flash floods, while approximately 5,000 homes were destroyed throughout KwaZulu-Natal. Roads and railway lines in the Durban metro, Eshowe, Pongola, Hluhluwe, Stanger and Port Shepstone, were damaged. Several rivers (Palmiet, Umdhlothi, Umlaas, Umbilo, Umhlathuzana, and Umhlangane) burst their banks and swept away six bridges on the 22nd of the month.

12/19–20   25 houses were flooded and 300 residents were left homeless when 95 mm of rainfall were recorded in Griekwastad (Northern Cape).

12/21–22 The Du Toit's Kloof Pass (Western Cape) was temporarily closed due to heavy rains, falling rocks, and mudslides. 131 houses were flooded in Calitzdorp. Apple, pear, peach, and grape orchards near Robertson, Ladismith, and Tulbagh were destroyed by hail and caused huge losses to farmers (approximately R20 million worth of damage in Tulbagh alone).

12/28 Severe thunderstorms caused extensive damage in the Marble Hall district (Gauteng). Many trees were uprooted and house roofs were ripped off.

January 2000

01/04–06 Heavy rains caused extensive damages to roads and destroyed vegetation and animals in several parts of the Northern Cape farming towns. 46 goats were drowned in these farming areas.

01/15–16 The Northern Cricket Union announced a loss of R2.5 million, due to the cancelled cricket match between South Africa and England. The cancellation was due to heavy rain conditions.

01/15–23 While heavy rain conditions were dominant over the summer rainfall region, more than 120 fires were fanned up by the strong southeasterly winds over the Cape Peninsula (Western Cape). The fire destroyed 12 of South Africa's best vineyards, natural vegetation, hundreds of township informal settlements and about eight houses. It was estimated that that the fire damage has reduced the grapevine exports by at least 25 percent. The worst fire damage was in the affluent suburb of Constantia, where damage was estimated at R20 million.

01/17–18 Several houses collapsed in Thohoyandou (Northern Province). At Guyuni, a four-year old girl died and her seven-year old sister was injured when a rain-soaked hut collapsed on them. The Makgobistad border post (between Botswana and South Africa) was forced to close temporarily for two days because of the flooding of the Molopo River. Estimates of R350 million in Mpumalanga and R200 million in the Northern Province were made for the repairs of bridges and roads.

February 2000

02/02 A severe storm with heavy rains and hail damaged 20 kilometers of land to the east of Newcastle (KwaZulu-Natal). Two people died in this event. Meanwhile, several houses, schools, businesses and motor vehicles were damaged by a severe hailstorm in Madadeni, Osizwenni, and Blaauwbosch.

02/05   A tropical low over Mozambique caused torrential rains over Northern Province and Mpumalanga.

02/23–24   The already overflowing rivers in Mpumalanga and Northern Province were further filled by rainfall which resulted from tropical cyclone "Eline" and caused extensive damage in Thohoyandou, Giyani, Messina, Louis Trichardt, Nzhelele, Tzaneen, Bushbuckridge, Graskop, and Nelspruit. Over 150 bridges and drainage structures were destroyed. At least 350 lives were lost in neighboring Mozambique and Zimbabwe. There were approximately 80 reported cases of death in the northern parts of South Africa, due to this natural disaster. Flood damage to the infrastructure of the Northern Province was estimated at R2 billion (rand). Meanwhile, an estimated 1 million people were treated for malaria and cholera in the Mpumalanga and Mozambique areas.

In summary, the anomalous weather conditions during 1998–99 and 1999–2000 have caused serious damage in most parts of South Africa. There have been many reported cases of death and injury due to the natural disasters, particularly in the less developed, informal settlements and rural areas. There has also been huge damage to the country's infrastructure, e.g., houses, roads, bridges, natural vegetation. The agricultural sector also suffered tremendously from irregular rainfall and tornadoes. The economic implication of this natural disaster negatively impacted the recovering economy, by increasing the inflation rate and other adversely affected economic indices.

ACKNOWLEDGMENTS

I would like to thank the members of the Research Group for Seasonal Climate Studies (SA Weather Bureau) who have assisted me with information for this project. Secondly, I would like to thank the personnel from the Climate Information Office and also the library for supplying me with the impact cases and newspaper clippings of these natural disasters.

REFERENCES

Cadet, D.L., 1985: The Southern Oscillation over the Indian Ocean. *Journal of Climatology*, 5, 189–212.

Lindesay, J.A., 1986: Relationship between the Southern Oscillation and atmospheric circulation changes over southern Africa, 1957 to 1982. Unpublished PhD Thesis, University of Witwatersrand, Johannesburg, South Africa.

Lutjeharms, J.R.E. and R.C. van Ballegooyen, 1988: The retroflection of the Agulhas current. *Journal of Physical Oceanography*, 18(11), 1570–83.

Jury, M.R. and B.M.R. Pathack, 1993: Composite climatic patterns associated with extreme modes of summer rainfall over Southern Africa: 1975–1984. *Theoretical and Applied Climatology*, 47, 137–45.

Tyson, P.D., 1986: *Climate Change and Variability in Southern Africa*. Cape Town (South Africa): Oxford University Press.

Asia and the Pacific

1998–99 La Niña in Indonesia: Forecasts and institutional responses

Kamal Kishore and A.R. Subbiah

Indonesian weather is impacted by east and west monsoon systems. The easterly wind system coincides with the dry season (April to September), and the westerly wind system coincides with the wet season (October to March). Generally, during El Niño, the easterly wind system is prolonged, causing a delay in the onset of the monsoon; during La Niña, the easterly wind system is weakened, resulting in the early onset of the monsoon and an enhanced distribution of rainfall.

While there are several studies that have focused attention on El Niño, no serious effort has been made to study the La Niña phenomenon's impacts in Indonesia. A preliminary study of La Niña in Indonesia reveals that in this century (1900–98), there were 19 La Niña years.[1] The salient features of monsoons during the past La Niña years in Indonesia were:

- Around 80 percent of the Indonesian archipelago received normal or above normal rainfall.
- The areas that received below normal rainfall were mostly agriculturally insignificant.
- The temporal distribution of rainfall during the wet season was either uniformly distributed over the monsoon season or in a few concentrated spells of heavy precipitation leading to localized floods.
- The early onset of the monsoon ensured a lengthy rainy season, normally starting from September–October. The termination of the rainy season either occurred as late as in June or abruptly in the first week of May, as happened in the Java region.

179

Table 3-8   Rice production in Indonesia (1994–98)

Year	Harvested area (ha)	Yield rate (q/ha)	Production (metric ton)	Annual growth (%)	ENSO state
1994	10,733,830	43.45	46,641,524	−3.20	El Niño
1995	11,438,764	43.49	49,744,140	6.65	Normal
1996	11,569,729	44.17	51,101,506	2.73	La Niña
1997	11,140,594	44.32	49,377,054	−3.37	El Niño
1998	10,680,898	43.34	46,290,461	−6.25	El Niño

Source: BPS (Central Bureau of Statistics, Indonesia).

This chapter focuses on the forecasts and institutional responses relating to the most recent La Niña during 1998–99. Before discussing the impacts of and responses to the 1998–99 La Niña, it is important to take into account the conditions that prevailed before its onset. The 1997–98 El Niño, which resulted in the late onset of the monsoon and reduced rainfall, coupled with long dry spells, significantly affected rice crop production. The details of rice production in 1998, compared to previous years, are shown in Table 3-8.

The reduction in rice production in 1998 by 5 million metric tons, when compared to the previous La Niña year of 1996, caused a serious threat to food security in the country. At the same time, the economic crisis adversely affected the country's import capacity, as reflected mainly by the declining availability of import credits from external suppliers. The interplay of these factors led to soaring food prices, rapidly rising unemployment, and a sharp reduction in access to food for a large segment of the population. It is estimated that 17 million families needed support against the spiraling prices of rice commodity.[2]

In May 1998, climate forecasts from IRI, NOAA, and Indonesia's Meteorology and Geophysics Agency (BMG)[3] indicated the possible demise of El Niño and the emergence of a La Niña in the remaining part of the year. There were apprehensions in Indonesia that La Niña would cause flooding, which would result in a loss of food grain production, further aggravating the food grain supply in the country.

Climate forecasts for 1998–99 La Niña and sectoral responses

Based on global indicators of an emerging La Niña, BMG issued a forecast on 4 September 1998 that in the 1998–99 wet season, almost all of Indonesia would experience high rainfall. The BMG representatives commented that the likely rainfall for each month would be on the order of 115 percent of the long-term mean rainfall for 102 meteorological districts. Climate-sensitive agencies like the Ministry of Agriculture and the Ministry of Water Resources perceived that La Niña would induce serious flooding. Media reports on La Niña also described it as a flood-inducing agent.[4]

In response to the forecasts, the climate-sensitive sector agencies prepared contingency plans. The information obtained from the Ministry of Water Resources and the Ministry of Agriculture is summarized below.

Ministry of Water Resources

The Ministry of Water Resources, on the basis of rainfall probabilities, identified 15 out of 27 provinces in the country as high flood-risk zones: Riau, South Sumatra, and Lampung provinces of Sumatra Island, all provinces of Java and Nusa Tengara regions, and South and East Kalimantan and Central Sulawesi provinces. These constitute around 60 percent of the country's territory. The Ministry of Water Resources established task forces in 15 provinces to prepare contingency plans to enhance preparedness and emergency planning efforts. Budgetary resources were augmented to the extent of 130 percent of the normal allocation, and all the task forces were authorized to invest in flood preparedness activities. A mechanism at the central government level to monitor disaster preparedness activities was established. This was the first time disaster preparedness activities were strengthened on the basis of the La Niña forecasts in August 1998, i.e., five months before the peak flood season.[5]

Ministry of Agriculture

The Ministry of Agriculture, based on historical data relating to flood impacts, categorized provinces and districts as high-risk, moderate-risk, and low-risk zones, and sent a communication to field agricultural extension agencies through the provincial departments to monitor these areas and undertake suitable measures in case of the occurrence of floods leading to crop damage.[6]

ADPC's pre-assessment of La Niña 1998–99 impact

In the wake of the 1998–99 La Niña, NOAA's Office of Global Programs commissioned the Asian Disaster Preparedness Center (ADPC) to undertake a rapid pre-assessment of the impacts of La Niña in Indonesia. In collaboration with NOAA and BAKORNAS PB of the government of Indonesia, ADPC fielded a mission to Indonesia to:

- Make a rapid assessment of the implications of La Niña 1998–99 for Indonesia, with focus on two main sectors of the economy: agriculture and food security, and natural resources and the environment; and
- Make broad, country-wide recommendations to mitigate the impact of the increased risk of disasters, and maximize potential benefits of early onset of rains and increased precipitation.

The mission analyzed historical data relating to La Niña impacts on Indonesia and found that it provided favorable climatic conditions for enhancing rice production by advancing the wet season and reducing the duration of the dry season. The mission also found that a maximum of only around 40,000 hectares out of the 11 million hectares of cropped area suffered total damage as a result of floods during past La Niña episodes. The mission, after discussions with farmers and experts, concluded that agricultural production would be favorable, and that flood damages were likely to be insignificant compared with the potential benefits. The mission recommended that rice production could be increased by adopting the following measures:

- Enhancing rice yield potential;
- Increasing crop intensity through restructuring the cropping pattern; and
- Expanding cropping area by putting idle fallow lands under rice cultivation.

The mission was of the view that the rice production level during 1998–99 could go up to the 1996 production level of 51.10 million metric tons compared to 46.29 million metric tons in 1997–98, if the following measures were to be undertaken:

- The La Niña forecasting information is incorporated into the country's agricultural development planning for the year 1998–99.
- An emergency rice production program is introduced by the government with adequate incentives for farmers to enable them to obtain agricultural inputs and take advantage of La Niña-induced increased precipitation.
- Short training programs are organized in a timely fashion to enable agricultural extension workers to use and disseminate La Niña forecast information for agricultural purposes. This will help in providing farmers with information on appropriate cropping pattern and agricultural inputs.
- As a follow-up to the above, location-specific contingency crop plans are prepared and disseminated to enable farmers to plant and implement appropriate cropping schedules.
- An effective coordination mechanism is established at each reservoir project level to regulate releases of water, keeping in view La Niña forecast information.

The mission noted that the government had a plan to import 3.2 million metric tons of rice for the fiscal year 1999–2000 (April 1999 to March 2000). The mission recommended that the government consider modifying the rice import plan, as it was prepared without taking into consideration the forecast of La Niña-associated climatic conditions in the 1998–99 wet season.

Actual impacts of the La Niña 1998–99

Flood incidence

Floods affected a total of 39,780 hectares of irrigated paddy fields, inundating some 5.5 km of national roads, 15.5 km of provincial roads, 60 km of local roads, and some 56 km of village roads. Water levels in the inundated area were recorded in the range of 15 cm to 2 meters. Six people were killed in flood incidents.[7] Some provinces, which were not included in the flood-risk zone, like South Sulawesi, were also affected by floods. Against the forecast that about 46 percent of Jakarta could be inundated by floods, the flood impact was actually confined to very few pockets.[8] The overall impact of flooding was insignificant. Hence, resources committed for flood preparedness proved to be disproportionate. The Ministry of Water Resources, with hindsight, was skeptical of utilizing the La Niña forecasts issued by the meteorology department (BMG) for operational decision making.

Crop production

In consonance with the pre-assessment of the ADPC mission, farmers in many parts of the country took advantage of the early onset of rain,[9] and the prevalence of high prices for rice motivated them to increase the area under rice cultivation above that of a normal year. Consequently, there was a significant increase in rice crop production.

In February 1999, because of advanced and enhanced harvests, the price of rice fell from around Rp 4,000 to around Rp 800 per kilogram.[10] Panic selling of rice by farmers at throw-away prices was reported. When the reports of distress sales came from many regions of Java, the government approved Rp 778 billion (about US$91 million) to provide credit to village cooperatives and non-governmental organizations for use in procuring rice from local farmers in order to stabilize the price of rice at Rp 1,520 per kilogram.[11] The government also decided to modify and reduce the import of rice to arrest further price drops.[12]

Lessons learned

Documentation and analysis of past extreme climate events and associated disaster events data could have led to a realistic pre-assessment and prevented the anticipation that 50 percent of the country would likely be flooded, as well as the commitment of huge resources for flood preparedness. Never during a past La Niña event had more than 40,000 hectares of area been affected by floods in Indonesia.

While farmers took advantage of early rains, thereby increasing rice production, government agencies could not take anticipatory action. Had the government based its actions on the La Niña forecast, it could have responded to farmers by providing resources to cooperatives to buy excess rice from farmers. The government also could have reduced its imports to prevent the flooding of imported rice onto the market.

The direct application of global ENSO parameters to local decision making poses serious difficulties. Policy planners and end-users were not able to utilize basic scientific information such as an SST index or the thermocline depth in the tropical Pacific for making resource management decisions. They judged the events in terms of socio-economic and environmental impacts that could be attributed to ENSO parameters in their particular area. Clearly, the local decision makers' interest is to know what are the specific impacts attributable to the precursor, growth, and decay phases of ENSO extremes on the southeast and northwest monsoons and in turn their impact on lowland and upland crops in a particular season.

Hence, there is a need to downscale and disaggregate the specific impact of ENSO parameters on monsoon patterns, and the impact of monsoon patterns on local climate and weather variables, and their impact on the given socio-economic system in a particular area. The disaggregation of ENSO-associated potential impacts on a temporal and a spatial basis would provide better resolution to climate forecast products, thereby enabling decision makers and end users to undertake proactive response measures.

Notes

1. ADPC and BAKORNAS PB Mission Report: *La Niña 1998–99: Challenges and Opportunities for Indonesia.*
2. FAO Mission Report, April 1998.
3. BMG Wet Season Forecast for 1998–99.
4. *The Jakarta Post*, 5 September 1998: Agency forecasts heavy rains, warns of floods. Heavy rains are expected to fall on the country from this month until early next year, the Meteorology and Geophysics Agency (BMG) said here on Friday. It warned of

possible floods and an adverse effect on harvests as a result of above-average precipitation, which is expected to last for between three and eight months depending on the region. The reason for the expected high rainfall, according to the agency's head Sri Diharto, could be the La Niña weather phenomenon, although, he added, it was still too early to be certain. "We will issue warnings once we have established if La Niña really has arrived," Diharto said in a news conference. Diharto said he hoped the relevant government agencies would begin to draft contingency plans for severe weather immediately. He said that public work offices should identify areas prone to flooding and help residents to prepare themselves and protect their crops and homes. He also said that agricultural offices should disseminate recommendations on crop planting patterns. "It's unlikely that this year's La Niña will be as strong, but *Allahu Akbar* (Allah is Great), who knows," he said.

    The Jakarta Post, 10 September 1998: Minister warns of La Niña-induced floods. The government has warned that La Niña-induced floods will threaten many Indonesian provinces, even as it admitted that the country has poor flood control contingency plans.

5. Personal discussion with Dr. Hafied A. Gany, Ph.D., Director, Directorate of Water Resources Management and Conservation (PSDA) in the Directorate General of Water Resources Development of the Ministry of Public Works, Government of Indonesia.
6. Personal discussion with Ir. Sutarto Alimoeso, and Mr. Yusmin, Agroclimatologist, Sub-Directorate of Crop Protection in the Directorate of Food Crops and Horticulture.
7. Paper presented by A. Hafied and A. Gany in the workshop organized by ADPC and BAKORNAS PB on Extreme Climate Events (ECE) in Indonesia, 11–12 February 1999.
8. Personal discussion with authorities of Directorate of Water Conservation and Management Division on 30 September 1999.
9. Paper presented by Ir. Sutarto Alimoeso, Director of Crop Protection, in the Workshop on Extreme Climate Events (ECE), organized by ADPC and BAKORNAS PB, 11–12 February 1998 in Indonesia.
10. Discussion with BULOG authorities, 10 February 1999 at Jakarta.
11. *The Jakarta Post*, 9 February 1999.
12. *Yahoo News Asia* (Internet), 5 March 1999: Indonesia Re-schedules Rice Imports to Ease Local Rice Price. The State Logistics Agency (BULOG) said on Thursday, it would re-schedule deliveries of rice to be imported this year to ease pressures on the price of locally produced rice. Industry and Trade Minister Rahardi Ramelan, who is also BULOG's chairman, said the agency would adjust the scheduling and destinations of the import deliveries to protect the local market. "In certain areas, rice delivery will be delayed until the closing of the harvesting season in April, and we will also unload some of the rice at areas that do not harvest this year," the *Jakarta Post* quoted Rahardi as saying. "The re-scheduling of the planned rice imports is needed to protect rice farmers from drastic falls in the price of rice in the local market," he said.

    The Indonesian Observer, 6 April 1999: RI to stop importing rice by 2000: Official. State Minister for Food and Horticulture A.M. Saefuddin said that Indonesia will no longer import rice by the year 2000. "If the natural condition remains like the current condition, then by February 2000, Indonesia will no longer need to import rice," he said, following a hand-over of cheap rice packages to fishermen in Ancol, North Jakarta yesterday. His prediction is based on the fact that the current domestic rice procurement has increased continuously during the present harvest time. In March this year, the realization of rice procurement is expected to reach over 450,000 metric tons, and in April is expected to reach the biggest amount of 600,000 metric tons.

La Niña and its impacts on China's climate

Wang Shao-wu and Wei Gao

The impacts of El Niño events on global and regional climate have been widely studied from the point of view of statistics and dynamics. However, the counterpart of El Niño – La Niña and its impacts – has not been paid much attention in China. In an El Niño year, the number of typhoon landings in China is greatly reduced. In Northeast China, there is a significant reduction of agricultural production in the event of a cool summer season.

El Niño has an impact on summer precipitation in China. The divergence caused by this impact is examined in this paper. In order to determine the impact of ENSO events on summer precipitation in China, the classification of El Niño events based on the geographical location or timing of the positive anomaly is redefined. The authors noticed that the redefinition of classifications could produce subjectivity in the analysis and therefore create difficulty in climate prediction. The relatively low significance of the relationship between ENSO and precipitation in China may be connected to the characteristics of ENSO events and other factors, such as the atmospheric circulation of the middle latitudes and sea surface temperatures in the "warm pool" of ocean water in the western Pacific Ocean. The latter has exhibited an independence to ENSO events, although it is negatively correlated with sea surface temperatures in the eastern Pacific Ocean. We did not use the stratification approach in this paper to examine the impacts on ENSO on the climate of China. Instead, the focus is on the impacts of La Niña events on precipitation in China.

Identification of ENSO events

Usually, ENSO events have been identified according to a single series of SOI or SST in the eastern equatorial Pacific, e.g., the SST in Niño3. However, uncertainty of any series increases in the earlier part of the record (such as in the late nineteenth century or early twentieth century). Thus, it is desirable to use a composite index instead of a single one. Wang and Shi (1992) indicated that the correlation coefficient was 0.65 between seasonal anomalies of SOI and SST in the Niño C region (0–10°S, 90–180°W) for the period 1871–1989. It infers that each of them can explain only less than one-half of the variance of the others; also a composite index conveys more information than a single index. Therefore, a composite ENSO index series was based on the following four indices:

- Niño3 SST (Kaplan et al., 1997, Cane et al., 1997). Spring 1856 to Autumn 1991. After 1991, the data set was updated using SST data from NOAA's CPC.
- Niño C SST (Angell, 1981). Summer 1867 to Winter 1987. After 1987 and gaps in the time series in the early period were updated, using the data of Wang and Shi (1992).
- SOI (Allan 1991; Ropelewski and Jones, 1987). Spring 1866 to Fall 1997. After 1997, NOAA's CPC data were used.
- SOI (Shi and Wang, 1989). The methodology used in construction of this series was the same as suggested by the CPC, but station data of sea level pressure (SLP) in the early years were different from those of Ropelewski and Jones (1987).

First, each of the four series was normalized by itself for the four seasons together in reference to the period of 1961–90. Then, the four normalized series were averaged to find a composite ENSO index series, for the period of spring 1867 to winter 1997. Before 1867, the series of the four indices were incomplete. In the making of the composite ENSO index series, positive (warm) SST anomalies were averaged with negative SOI anomalies. Then, a positive composite index corresponds to positive phase of SST and negative phase of SOI, and vice versa. Finally, La Niña events (negative [La Niña] composite index) for the period of 1867–1997 were identified, when the composite index was −0.5 and persisted at least two seasons.

Table 3-9 Relationship between La Niña events and climate in China

La Niña	I-1	I-2	I-3	I-4	I-5	I-6
1880					/	/
1882					S	S
1886			S		S	
1887	S		S	S		
1889			S			
1890			S		S	
1892	S				S	
1893	S		S	S	S	S
1894	/	/	/	/	S	S
1898					S	
1903			S		S	S
1906			S	S		S
1908						
1909	S		S	S		S
1910	S			S		S
1916		S		S		S
1917		S	S			S
1921			S			S
1922			S			S
1924	S	S				
1933	S	S		S	S	S
1937	/	/	/	/		
1938		S	S	S		S
1947			S		/	/
1950		S	S		S	S
1954			S	S	S	S
1955		S	S			
1956			S	S	/	/
1964				S	/	/
1970		S		S		
1971	S				S	
1973		S	S			S
1974	S				/	/
1975		S		S	S	S
1988				S	S	
F	9/33	10/33	18/33	14/33	14/29	17/29
r	−0.10	−0.25	−0.36	0.26*	−0.23*	0.48*

S indicates I-1 $\geq$ 9, I-2 $\geq$ 0.5 °C; I-3 $\geq$ 10%, I-4 $\leq$ −10%, I-5 $\geq$ 25%, I-6 $\leq$ −20%.
* Indicates correlation coefficient is significant at 95% of confidence level.

Impacts of La Niña events

Thirty-two La Niña events were found for the period of 1867–1997, and are listed in Table 3-9. Impacts of La Niña on the climate of China were investigated by using the following six climatic indices:

I-1. The number of landed typhoons.

I-2. Summer temperature anomalies averaged for 10 stations covering East Asia between 40°–53°N, 120°–145°E.

I-3. Summer precipitation anomalies averaged for five stations in north and west China (Beijing, Zhengzhou, Yichang, Chengdu, and Chongqing).

I-4. Summer precipitation anomalies averaged for five stations in south China (Fuzhou, Shantou, Guangzhou, Zhanjiang, and Nanning).

I-5. Winter precipitation anomalies averaged for five stations in northwest China (Taiyuan, Zhengzhou, Xian, Yinchuan, and Lanzhou).

I-6. Winter precipitation anomalies averaged for five stations in southeast China (Wenzhou, Fuzhou, Jian, Guangzhou, and Guilin).

Table 3-9 indicates that, when La Niña events occurred, precipitation increased in the north and west, and decreased in the south in summer, and increased in the northwest and decreased in the southeast in winter. Meanwhile, typhoon landings increased, and summer temperatures over East Asia were higher than normal. However, the correlations to ENSO for the latter two indices were weaker than the others. Correlation coefficients calculated by using the whole time series for 1880–1997 are given in the last line. Letter "S" shows the accordance of the indices with La Niña events. "F" indicates the number of accordances over the total number.

REFERENCES

Allan, R.J. et al., 1991: A further extension of the Tahiti-Darwin SOI, early ENSO events and Darwin pressure. *Journal of Climate*, 4(4), 743–9.

Angell, J.K., 1981: Comparison of variations in atmospheric quantities with sea surface temperature variations in the equatorial eastern Pacific. *Monthly Weather Review*, 109(2), 230–41.

Cane, M.A. et al., 1997: Twentieth-century sea surface temperature trends. *Science*, 275, 957–60.

Kaplan, A. et al., 1997: Reduced space optimal analysis for historical datasets: 136 years of Atlantic sea surface temperatures. *Journal of Geophysical Research*, 102(13), 27,835–60.

Ropelewski, C.F. and P.D. Jones, 1987: An extension of the Tahiti-Darwin Southern Oscillation Index. *Monthly Weather Review*, 115(10), 2161–5.

Shi, W. and S.-W. Wang, 1989: SOI, 1857–1987. *Meteorology*, 15(5), 29–33 (in Chinese).

Wang, S.-W. and W. Shi, 1992: Impacts of two kinds of ENSO on summer rainfall in China: *Proceedings of Studies on the Impacts of Oceans*. National Oceanic Administration, Beijing: Ocean Press, 76–87 (in Chinese).

Impact of the 1998 La Niña on Indian monsoon rainfall

R.H. Kripalani and Ashwini Kulkarni

Seasonal variation of rainfall is the most distinguishing feature of the monsoonal regions of the world. About 80 percent of the annual rainfall over a large part of India occurs during the summer monsoon period (June to September). The mean Indian monsoon rainfall (IMR) for the country as a whole is 852 mm, based on the long data period 1871–1999 with a standard deviation of 84 mm. The IMR is defined as normal when its value falls within plus or minus 10 percent of its mean value. If the IMR for a particular year is less (more) than 90 percent (110 percent) of the mean it is defined as a deficient (excess) monsoon. The year-to-year variability in monsoon rainfall occasionally leads to extreme hydrological events (i.e., large-scale droughts and floods) over different parts of the country, resulting in a serious reduction in agricultural output and affecting the national economy. In view of the critical influence of such variability, the forecasting of monsoon rainfall assumes profound importance for policy making and planning of mitigative effects. One of the factors causing these extreme hydrological events is the El Niño/Southern Oscillation (ENSO) phenomenon.

ENSO and Indian monsoon rainfall (IMR)

It is now well recognized that the ENSO phenomenon is one of the most important modes of the earth's year-to-year climatic variability. In gen-

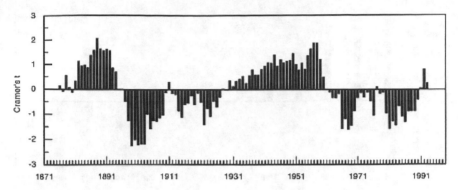

Fig. 3-26 Values of Cramer's t-statistic for the 11-year running means depicting decadal variability in Indian monsoon rainfall (data period: 1871–1999).

eral it is believed that the warm phase (i.e., El Niño) of the ENSO phenomenon is associated with the weakening of the Indian monsoon with an overall reduction in rainfall, while the cold phase (i.e., La Niña) is associated with a strengthening of the Indian monsoon. However, recent analysis for a 129-year period (1871–1999) has fostered new insight into these teleconnections.

While the year-to-year IMR series exhibit random fluctuations, a comparison of decadal IMR with the overall mean shows interesting features. Figure 3-26 shows the values of the Cramer's t-test for the 11-year (decadal) running means. This statistic compares the means of the subperiods (here 11 years) with the mean of the whole period, and isolates periods of above and below average rainfall.

The most striking feature seen in this figure are the epochs of above and below normal rainfall. The periods 1880–95 and 1930–63 are characterized by above normal rainfall, while the periods 1895–1930 and 1963–90 are characterized by below normal rainfall. Further analysis reveals that the impact of El Niño (La Niña) on IMR is more severe during the below (above) normal epochs than the above (below) epochs. Thus the impacts of ENSO extreme events are modulated by the decadal scale behavior of IMR. A close examination of Figure 3-26 reveals that the IMR entered into an above normal epoch around 1990. This may be a possible reason for the weakening of the El Niño-IMR relationship after 1990. None of the El Niños after 1990 have had any adverse effect on IMR. For example, in spite of the severe 1997 El Niño, the IMR was 102 percent of the long-term mean.

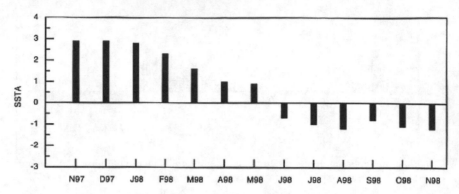

Fig. 3-27 Monthly sea surface temperature anomalies (SSTA) over the Niño3.4 (5°N–5°S, 170°W–120°W) region, November 1997–November 1998 (from NOAA/NWS/NCEP, *Climate Diagnostics Bulletin*, No. 98/11).

La Niña of 1998

The recent severe 1997–98 "El Niño of the Century" reached its maximum peak around December 1997, when sea surface temperature (SST) anomalies (the actual temperature minus the mean temperature) were at a maximum. Thereafter, SSTs continued to decrease, and after May 1998, the decrease was sharp. Figure 3-27 illustrates this point. It depicts SST anomalies from November 1997 through November 1998 for the region designated as Niño3.4 (5°N–5°S, 170°W–120°W). It shows that the temperatures declined and, after June 1998, moderate cold episode conditions (La Niña) commenced. This weakening of the warm episode (El Niño) and the entry of the cold episode (La Niña) released a huge amount of heat from the tropical Pacific. The effects of this heat were carried around the globe by atmospheric circulation. This contributed to, if not led to, the unprecedented heat-wave conditions across Asia during May and June 1998.

During July–September 1998, after the heat wave, monsoon floods left a trail of death and destruction in their wake. Monsoon rains played havoc in parts of northern India, particularly the northeastern region. The world-famous Kaziranga National Park for animals was submerged under 18 feet of water, leaving no animals behind in the 430 square kilometers of the park. Floods in July and August 1998 inundated three-quarters of Bangladesh – the worst flooding in Bangladesh's history. China also experienced its worst flooding in decades. The flood peak in the Yangtze River was the second largest since 1954. In South Korea, flooding was caused by the heaviest August rains in 27 years. In Japan, the heaviest rain in more than 80 years (e.g., since 1914) deluged the

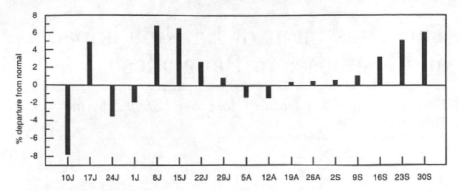

Fig. 3-28 Weekly cumulative area-weighted Indian monsoon rainfall (percentage departure from the long-term mean) for the week ending 10 June through 30 September 1998 (India Meteorological Department).

north-central part of the country. Borneo also reported floods after the torrential rains. Indonesia, which had suffered devastating droughts and forest fires because of El Niño, was hit by heavy rains and floods. The entry of La Niña may have triggered these floods (see Kishore and Subbiah, this volume).

The Indian monsoon rainfall was normal up to the end of July 1998. After that, the cumulative area-weighted rainfall for India showed a monotonic increase. Figure 3-28 illustrates this point; it shows the cumulative rainfall, at the end of each week, as a percentage departure from normal. The total seasonal IMR for the period from June to September 1998 was 106 percent of the long-term average (905 mm, compared to the normal 852 mm). The tropical Pacific entered into a La Niña phase after June 1998 (Figure 3-27) and the IMR showed a monotonic increase in rainfall after August 1998 (Figure 3-28). Since the IMR was in the above normal epoch (Figure 3-26), the entry of La Niña (Figure 3-27) was conducive for good monsoon activity over India (Figure 3-27).

Summary

The Indian Monsoon Rainfall regime entered into an above normal epoch around 1990. This implies that the IMR may be affected by La Niña but not necessarily by El Niño during the next decade or so.

The assessment of La Niña impacts and responses in Bangladesh

Ekram Hossain, Kzi Hasan Imam, Syed Shamsul Alam, and Monjurul Hoque

Although Bangladesh is one of the worst victims of the impacts of both El Niño and La Niña, La Niña features are considered to be more predominant. The existing geophysical and socio-economic settings of the country help to increase both the vulnerability to and severity of the impacts. Scientists have found a correlation between La Niña events and the variability of climatic phenomena in the country which, as a result, cause climate-related natural disasters like cyclones, floods, nor'westers, tornadoes, etc. To reduce the negative impacts of those climate-related disasters and to minimize the sufferings of the people, the government of Bangladesh established a set of mechanisms including institutional arrangements both at national and local levels for disaster preparedness, and relief and rehabilitation of the areas affected or likely to be affected.

Many scientists around the world still refer to the impacts of the extremes of the El Niño-La Niña process (i.e., the ENSO cycle) as one of symmetry. Our experience, however, suggests that the opposite impacts of about equal strength and intensity do not always accompany each type of ENSO extreme. Much about the process is still unknown. Many of the local or regional factors that tend to increase or decrease the severity and intensity of the ENSO extremes are yet to be uncovered and then used in predicting and analyzing the behavioral patterns of such events.

Scientific research in Bangladesh relating to El Niño has not yet reached a satisfactory level, and studies of La Niña are only just beginning. Research on ENSO is mostly conducted as a result of individual initiatives. There is, however, some evidence of historical interest about La Niña in the country. The real credit for meteorological and agro-meteorological predictions during ancient times goes to a mythical woman named Khana. Her verses are said to be the envy of any scientist of any time. The following examples were attributed to her many centuries ago:

(i) "During the ninth lunar day of the bright fortnight in the Bengali month of *Ashar* [June–July], if the rain is torrential, the year may experience drought; if rainfall is intermittent, the year may experience heavy flood; if moderate rainfall occurs, crops will grow in abundance and if the sky remains clear and bright during sunset on that day, the year will experience salvation"; or
(ii) "If the Northern wind blows during the month of *Shraban* (July–August) the year will experience severe floods."

Today's experts have found a scientific basis behind these predictions. Scientifically, Gilbert Walker, during the British rule in India, identified what are now considered ENSO effects on this sub-continent.

La Niña teleconnections

The scientific views about the existence and strength of La Niña teleconnections to Bangladesh have been best explained by Walker's observations during the 1920s. In Bangladesh La Niña corresponds to the positive value of the Southern Oscillation Index (SOI) and results in increased rainfall leading mostly to deluges. According to his observations, low atmospheric pressure generally prevails in the region from Australia to India, when high atmospheric pressure prevails in the eastern tropical Pacific Ocean. The scientific reason behind this is that wind flows from a high-pressure region to a low-pressure region. As a result, a huge volume of moisture comes from the Pacific Ocean to Bangladesh and India. This is due to prevailing low atmospheric pressure in this area at the time that moisture increases, causing heavy rainfall in this region. The present study reveals that La Niña results in increased rainfall, floods, and cyclones. The time-series data of yearly rainfall of four selected recording stations of Bangladesh for a period of 43 years (1950–92) show a negative and decreasing tendency in rainfall during El Niño events.

The study reveals that during major El Niño years, at least during the first year of El Niño, Bangladesh can be spared from catastrophic floods, and during the La Niña (positive SOI) and weak El Niño years, Bangladesh can be a victim of flood. In the case of high positive SOI Bangladesh may face severe floods. Statistical evidence shows that the most catastrophic floods in Bangladesh occurred in 1954, 1955, 1974, 1987, 1988, and in 1998. The years 1954, 1955, 1988, and 1998 were years with positive ENSO indices while 1987 was a year of El Niño. The main El Niño occurred in the previous years and in these years the negative anomalies were not so strong. In major El Niño years, i.e., 1951, 1957, 1972, 1976, 1982, and 1986, Bangladesh did not experience a catastrophic flood. Thus one can come up with the hypothesis that during major El Niño years, at least during the first year of El Niño (its onset year), Bangladesh could be spared from catastrophic flooding. The study also shows that when the SOI is small (positive or negative) and when the 28.5 °C isotherm stays left of 165°E longitude, the chance that Bangladesh could be hit by a cyclone is quite high.

A significant amount of excessive rainfall in the selected stations under study was observed in the preceding and following years of the 1982–83 El Niño event. It is evident from the study that Bangladesh was affected by the 1997–98 El Niño and subsequently by the 1998–2000 La Niña. Moreover, unprecedented and unusual foggy weather during the winter and high temperatures during the southwest monsoon of 1998 were observed. The subsequent La Niña impact after the 1997–98 El Niño resulted in a severe and prolonged flood in Bangladesh in 1998 where about 51 percent of the total area of the country was inundated and approximately 31 million people (26 percent of the total population) were negatively affected. The total losses as a percentage of the country's GDP amounted to 6.64 percent. This drop in GDP negatively influenced the economic growth rate of the country.

During strong El Niño years, it has been observed that Bangladesh has not been affected by catastrophic floods. The chance of Bangladesh being hit by a cyclone is higher when the SOI is strongly positive (La Niña conditions). Again, Bangladesh was spared from catastrophic cyclones when the SOI was strongly negative or the 28.5 °C isotherm lay east of 160°W longitude. When the SOI is positive, the Walker circulation is strong, atmospheric pressure in the southeastern Pacific is high, and in the Australian region it is low and the winds are easterly (westward flowing). However, when the SOI is negative, the Walker circulation is weakened and the easterly winds are weakened and sometimes reversed. When the Walker circulation weakens, the Hadley circulation gets stronger and the meridional (north-south) winds get stronger. Scientists believe that the strength of the Bangladesh monsoon is governed by the movement of tropical disturbances that form in the Pacific. The Bay of Bengal during the monsoon period, which is the origin of most of the rain in Bangladesh and the catchment areas of rivers, have their origin in most of the depressions or atmospheric disturbances formed in the Pacific. When the SOI is positive and high, the Walker circulation is strong, the upper tropospheric wind in the Australian region is easterly and, consequently, the tropical disturbances are transported westward, moving into the Bay of Bengal. If El Niño is weak or moderate, the Hadley circulation may not be very strong, which allows some of the tropical disturbances to cross into the Bay of Bengal and enter into Bangladesh causing floods and cyclones, as Bangladesh falls between 22–26°N of the Equator.

In order to understand the monsoon phenomenon, we cannot confine ourselves to the Indian Ocean alone. The control of the monsoon seems to lie in the Pacific Ocean and it is the Walker circulation that largely governs monsoon variability in Bangladesh.

La Niña has many impacts on the economy and society of Bangladesh. Floods and cyclones destroy the natural, physical, economic, and socio-cultural settings to a large extent. Damage in agricultural production, increased marginalization of the population, increases in the instances of diseases and hunger, increased salinity in water supplies, mass rural-urban migration, infrastructure damages, etc. are among the major La Niña impacts in the country. These impacts result in economic disorder, social unrest, and increased criminal activities in the country.

Response strategies

The development of response strategies to La Niña's impacts in Bangladesh is still in its infancy. "Development of Models for Predicting Long Term Climate Variability and Consequent Crop Production as Affected by El Niño-La Niña Phenomena," undertaken by SPARRSO (Space Research and Remote Sensing Organization) and BARC (Bangladesh Agricultural Research Council), is the only notable research.

Governmental responses to La Niña-spawned hazards, like floods and cyclones, include (i) the broadcasting of advance warning and continuous monitoring of the situation, (ii) the creation of local and national level controls and monitoring cells for flood situations, and the undertaking of required relief and rehabilitation programs, (iii) the ensuring of mass participation in extending help and cooperation to the victims, (iv) the seeking of help from all sources to strengthen the relief and rehabilitation programs undertaken by the government.

Forecasting by analogy

Some scientists have suggested that forecasting the onset of La Niña is a more difficult task than forecasting the onset of El Niño (see Zebiak, this volume). Bangladesh has not yet developed powerful institutions like those that exist in Western Europe and North America to forecast a complex climate-related phenomenon like El Niño or La Niña, but the institutions established at the national level such as Space Research and Remote Sensing Organization (SPARRSO), the Bangladesh Meteorological Department (BMD) and the Bangladesh Water Development Board (BWDB) are trying their best to monitor and forecast climate-related hazards in order to avoid the surprising emergence of disasters.

Because the present level of technology in Bangladesh is not yet able to enable its researchers to unfold the mysterious impacts of La Niña in Bangladesh, the country has to depend on the latest ideas and technologies from the developed countries in order to challenge the negative impacts of La Niña events. Regular exchanges of information with specialized organizations in the developed countries and a proper understanding of the mechanisms of the ENSO warm event/cold event cycle would enable researchers in the country to better forecast, plan, and monitor the country's natural hazards like cyclones, tornadoes, nor'westers, and floods, that are often spawned by ENSO's extremes.

Policy recommendations

The major recommendations of our study are as follows:
* A clear and comprehensive national disaster policy that addresses the total disaster management spectrum, including El Niño/La Niña and the consideration of all aspects of preparedness and mitigation, needs to be formulated.
* A separate cell to be named as "El Niño/La Niña forecasting cell" should be created to integrate and coordinate the El Niño/La Niña-related activities of the concerned agencies. Since El Niño/La Niña impacts severely affect the intensity of the monsoon destroying the harmony of nature, the cell should be well equipped to conduct intensive studies on various El Niño/La Niña issues and their interrelations with those caused by the monsoon.
* Since La Niña impacts are more predominant than those of El Niño impacts in Bangladesh, emphasis should be given more to La Niña studies.
* Specific issue-based scientific research on El Niño/La Niña impacts should be carried out through international collaboration.
* The concerned scientists should be properly trained in predicting and monitoring El Niño/La Niña impacts.
* A short, medium, and long-term Disaster Management Action Plan, including El Niño/La Niña considerations should be developed within SAARC (South Asian Association for Regional Cooperation) or other regional framework for cooperation.

The influence of La Niña on the rainfall of the Western Pacific

Charles "Chip" Guard

The isolated nature of tropical and subtropical Pacific islands makes the local availability of fresh water critical. Under normal conditions, most Pacific islands have ample fresh water supplies from a number of sources. Mountain islands normally have streams and rivers that provide a source of surface water that is commonly used for drinking, cooking, bathing, and washing clothes. Larger mountain islands may even have lakes. Mountain islands also have subsurface sources of water in the form of a fresh water lens that floats atop salt water beneath the limestone (coral) parts of the island. These are the more coastal parts of mountain islands. Rainwater flows along the impervious basalt until it reaches the porous limestone, then flows underground toward the ocean. Well-placed wells can tap this fresh water lens. However, on many mountain islands, the subsurface water resources are not well developed due to the normal abundance of cheaper, more accessible surface water.

Atolls are totally composed of porous carbonates, and thus there are no persistent sources of surface water. However, these atolls also have a fresh water lens that floats atop the salt water. The larger islets of atolls can have a substantial fresh water lens, but small islets can have only a small, relatively thin lens. Still, the fresh water can be tapped with shallow wells. If the pumping is too strong or too deep, it can damage the delicate interface between the fresh water lens and the salt water, producing brackish water. Once damaged, the interface cannot restore itself until the rains return. Restoration of the lens can take months. Another common source of fresh water used on atolls is the direct catchment of rainwater. Rainwater is often captured from roof tops and funneled into storage tanks.

Most Pacific tropical islands (within 8–10 degrees latitude of the equator) normally receive abundant rainfall year round. The more subtropical islands (8–10 degrees latitude to 15–20 degrees latitude) experience distinct wet and dry seasons. North of the equator, the wet season is generally from July through October and the dry season is generally from January through April. The other months are transition months, which may be wet or dry depending on the synoptic weather patterns. In the Southern Hemisphere, the wet season is January through April and the dry season is July through October.

Occasionally, the Pacific islands can experience extended periods of excessive rain and extended droughts. It is during these periods of weather extremes that islands frequently run into trouble. As we have discovered in recent years, most of these extremes are associated with El Niño and La Niña events. For example, during an El Niño onset (March to September), Indonesia and Southeast Asia experience drought. South-central Micronesia (Chuuk, Pohnpei, Kosrae, Majuro), Papua New Guinea, western Kiribati, Nauru, and Tuvalu experience strong westerly winds and persistent heavy rains that can cause flooding and landslides. During the mature stage of El Niño (October to May), drought moves into Micronesia, Papua New Guinea, western Kiribati, Nauru, and Tuvalu. The strong westerly winds and heavy rains propagate eastward to eastern Kiribati, Tokelau, Samoa, the northern Cook Islands, and northern French Polynesia. Normally rare tropical cyclone activity occurs near Hawaii, Samoa, the Cook Islands, and French Polynesia.

During the first half of La Niña (July–December), heavy rains return to Indonesia and Southeast Asia and affect subtropical Southern Hemisphere regions (New Caledonia, Fiji, Tonga, Niue, the southern Cook Islands, and southern French Polynesia). Drought affects Nauru, all of near equatorial Kiribati, Tuvalu, Tokelau, Samoa, the northern Cook Islands, and northern French Polynesia. During the second half of La Niña (January to June), heavy rains affect extreme northwestern Kiribati and Micronesia from 4°N to 10°N. During this time, drought continues along the equator from the eastern Pacific to 160°E (Kiribati and Nauru) and also continues to affect Tuvalu and Tokelau.

The occurrence of heavy rains and strong westerly winds during El Niño can cause flooding and ocean inundation. The persistent high surf occasionally takes the lives of children and fishermen. The strong westerly winds can cause inundation to occur on the lagoon side of atolls where people often build their houses. The biggest threat on mountain islands is the threat of landslides. In March 1997, an El Niño-enhanced rainfall event caused over 20 inches of rain to fall on Pohnpei in 24 hours. Already saturated, the mountain slopes in one area could no longer adhere to the underlying rock. A portion of the mountainside collapsed, burying a small fishing village. Nineteen people died in the few seconds of the landslide. However, most islands can handle most heavy rain events.

For the Pacific islands, drought is potentially the most devastating event. First of all, drought sneaks up on the population. The apparent changes in rainfall that precede drought are often subtle. Rain still occurs, but it begins to be less frequent, less intense, and less in duration. Then suddenly, it stops. By the time the drought becomes apparent, it is too late to take actions that could have conserved the water resources. Thus, response to a drought must occur based on the prediction of the drought several months ahead of time – at a time when heavy rains may still be falling.

Drought produces several impacts. On mountain islands, it causes the surface water sources to diminish or to dry up. Stream and river flow is reduced, the water becomes stagnant, and gastro-intestinal diseases flourish. Diarrhea causes dehydration. During a drought, this occurs when the temperature is hotter and available drinking water is scarce. On mountain islands, the problem is often increased inconvenience rather than not having any water at all; that is, it is more difficult to access the available water. Drought does have an impact on the vegetation and agriculture of mountain islands. The hot dry weather causes the normally damp land to become very susceptible to grass and forest fires. Frequently, these fires cannot be controlled or put out because there is no source of fresh water. During the 1997–98 El Niño-related drought, 20 percent of Pohnpei experienced fires. During the same period, Guam recorded some 1,400 grass fires.

On atolls, drought can be a life-threatening event. After the drought begins, there is no more water to recharge the fresh water lens. The fresh water in the lens continues to drain into the ocean, and the lens becomes thinner and thinner. Eventually, the water becomes brackish, then nonpotable. The brackish water also affects agriculture, and the taro and tapioca die, coconuts shrivel up and breadfruit stops production. On atolls, along with the drought comes famine. Even after the rains return, it takes 8–12 months for the crops to recover. Many trees die and do not recover. With severe water rationing, poor water quality, and shriveled coconuts, dehydration becomes a serious problem. Cloudless skies exacerbate the situation. Under British rule, in past droughts, the British evacuated the total population from atolls.

La Niña occurs when the equatorial (5°N–5°S) eastern and central Pacific monthly sea surface temperature (SST) anomaly is more than 0.5 °C below the normal average monthly SST for a continuous period of four

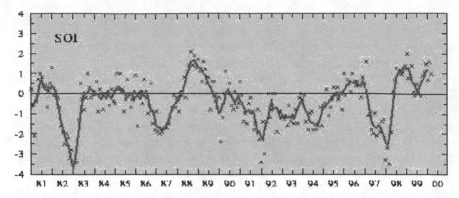

Fig. 3-30 Five-month running mean of the Southern Oscillation Index (SOI) since 1980. Individual monthly values are indicated by x (from the *Climate Diagnostics Bulletin*).

months (Trenberth, 1997). We frequently use the average monthly SST in the Niño3.4 area (170°W–120°W) as the baseline. The cooler SSTs produce a characteristic tongue of cool water anomalies along the equator (see Figure 3-29 (*c*)). These characteristic cooler waters are seen in Figure 1-2, which shows the major La Niña events that have occurred since 1949. The blue shaded areas coincide with La Niña events. The waters generally begin to cool in June or July and reach their coldest values from November to January.

This oceanic behavior is closely associated with an atmospheric response, which can be measured by positive values of the Southern Oscillation Index (SOI) – a normalized value of the pressure anomaly at Tahiti minus that at Darwin, Australia. Five-month running means of the SOI are shown in Figure 3-30. Positive values (above the center line) are generally associated with La Niña events. Accompanying changes in the SOI are changes in the low level and upper level equatorial winds, in the adjacent vertical motion fields, and in the cloud and rainfall distributions. Table 3-10 shows the Niño 3.4 SST anomalies, the SOI values, the wind field anomalies in the western and central equatorial Pacific, and the rainfall values and percent of normal rainfall at Pohnpei during the La Niña cycle.

During La Niña, the equatorial-low level easterly winds (westward flowing) are stronger than normal in the western and central Pacific. Associations are not so clear-cut in the equatorial eastern Pacific. Upper level westerlies in the central Pacific are also stronger than normal. Positive values of the outgoing longwave radiation (OLR) indicate that the equatorial central Pacific is clearer (less cloudy) than normal during La Niña.

Table 3-10 Some parameters used to monitor the ENSO cycle

1	2	3	4	5	6	7	8	9	10
			850-hPa wind index	850-hPa wind index	850-hPa wind index	200-hPa wind	OLR	mo	mo avg
mo/yr	SST	SOI	WP	CP	EP	CP-EP	DL	rain	rain
Oct/98	−1.1	+1.0	1.6	1.2	0.4	0.3	1.2	14.81	16.71
Nov/98	−1.2	+1.1	2.1	1.4	0.1	1.0	1.7	17.35	15.74
Dec/98	−1.6	+1.4	1.9	1.7	0.4	0.7	2.0	19.91	15.22
Jan/99	−1.5	+2.0	3.4	1.6	0.1	2.2	2.3	18.22	13.07
Feb/99	−1.2	+0.8	1.4	1.4	−1.0	1.7	1.6	18.05	10.80
Mar/99	−0.8	+0.9	3.5	1.2	−1.1	0.0	2.0	27.33	13.54
Apr/99	−0.7	+1.4	2.9	1.0	−0.8	0.6	1.3	21.20	16.44
May/99	−0.6	+0.1	1.8	0.8	−0.4	−0.3	0.9	12.13	19.12
Jun/99	−0.8	−0.1	1.0	0.3	−1.0	1.2	0.7	11.55	17.14

Columns 2 and 4–8 are values between 5N and 5S. SST is the Niño3.4 average sea surface temperature (170W–120W). Wind index represents deviation from climatological monthly mean; WP = 135E–180, CP = 175W–140W, EP = 135W–120W, and CP-EP = 165W–110W. For 850 hPa wind indices, positive (negative) values imply easterly (westerly) anomalies; for 200 hPa wind indices, positive (negative) values imply westerly (easterly) anomalies. OLR is the outgoing longwave radiation (positive values indicate below normal cloudiness) and DL = 165E–165W. Rainfall is in inches and average rainfall is from the 1961–90 30-year average. (Columns 2–8 from the *Climate Diagnostics Bulletin*; rainfall from the Weather Service Office at Pohnpei).

In the Northwest Pacific, La Niña is often associated with much heavier than normal rainfall between 3°N and 10°N, primarily during the dry-season (December–May). The strong Northern and Southern Hemisphere trade winds associated with La Niña converge to produce a strong trade-wind trough in the eastern half of Micronesia between 3°N and 8°N. Tropical disturbances develop in the trough and propagate to the west or west-northwest, producing very heavy rainfall in their paths. The disturbances that develop in the trade wind trough occasionally interact with shearlines (washed-out cold fronts), take a more northwestward track, and move over Yap and the Mariana Islands, producing heavy rainfall there. The equatorial areas of the western Pacific, east of 160°E, are very dry during this period. The heavy La Niña rainfall in Micronesia is generally confined to a narrow band from 3°N to 9°N, with very large rainfall gradients.

There can also be considerable rainfall variability among La Niña events. This is illustrated in Figure 3-31, which shows the differences between the rainfall in Micronesia during the La Niña of 1998–99 and the La Niña of 1999–2000. During these cold events, tropical cyclone activity

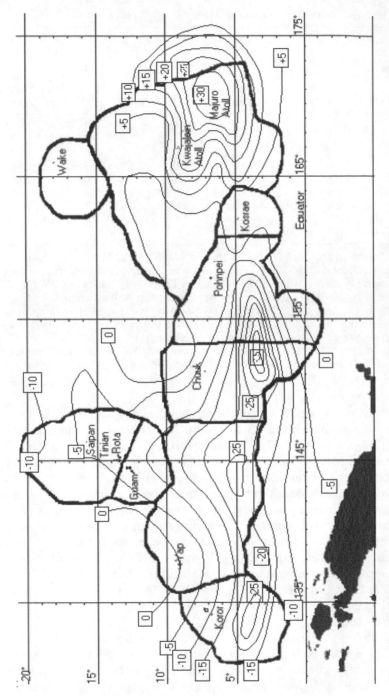

Fig. 3-31 January–March 2000 rainfall in Micronesia minus January–March 1999 rainfall in inches. Greatest departures are in the southern Marshall Islands (right edge of picture), southwest of Pohnpei (center of picture), and over Palau (left edge of picture). Heavy lines delineate national, state, or territorial boundaries (from the *Pacific ENSO Update*).

Table 3-11 Annual rainfall amounts (inches) and percent of normal annual rainfall for 1999 at selected Micronesian Islands and for Pago Pago, American Samoa

Location	Lat/Long	1999 Rainfall (in)	% Normal	2000 Rainfall Jan–Mar (in)	% Normal
Pago Pago, American Samoa	14.3S/170.7W	126.79	104	45.21	125
Guam Int'l Airport, Guam	13.5N/144.8E	86.87	95	11.76	110
Andersen AFB, Guam	13.6N/144.9E	96.02	98	14.36	99
Rota, CNMI	14.1N/145.2E	92.03	97	10.39	109
Saipan Int'l Airport, CNMI	15.1N/145.7E	68.15	92	6.59	76
Capitol Hill, Saipan	15.2N/145.8E	82.27	99	12.39	139
Tinian, CNMI	15.0N/145.6E	71.59	93	9.67	109
Chuuk, FSM	7.4N/151.7E	154.78	115	34.98	151
Pohnpei, FSM	6.8N/158.1E	194.65	103	59.81	161
Nukuoro, Pohnpei, FSM	4.0N/155.0E	225.06	151	44.44	129
Pingalap, Pohnpei, FSM	6.5N/160.5E	220.33	124	65.66	167
Kapingamwarangi, Pohnpei, FSM	1.7N/154.8E	65.46 (4/99-3/00)	68	9.83	28
Kosrae, FSM	5.3N/163.0E	214.51	104	61.85	141
Koror, Palau	7.3N/134.5E	171.89	116	39.96	142
Majuro, RMI	7.5N/171.4E	128.17	98	51.35	234
Kwajalein, RMI	8.8N/167.7E	86.36	85	25.42	213

Continuous shaded rows indicate common geographical areas (top – Mariana Islands; center – Pohnpei State; bottom – Marshall Islands).

and monsoon activity were delayed and were pushed to the north and west away from Micronesia. As a result, the rainy season is frequently drier than normal. The very wet dry season and the somewhat drier than normal rainy season combine to make the total yearly rainfall nearly normal. However, La Niña-associated rainfall can be so overpowering as to make the total annual rainfall well above normal at some locations. This can be seen for Nukuoro and Pingalap in Table 3-11.

The Southern Hemisphere responds somewhat differently to La Niña than does the Northern Hemisphere. In the Southern Hemisphere, the strong easterly trade winds also shift the monsoon and tropical cyclone activity to the west, toward the coast of Australia. Near-equatorial areas to 10°S and east of 170°E are subjected to strong subsidence and drought occurs in Kiribati, Tuvalu, Tokelau, Niue, and the southern Cook Islands. A cross-section of the atmosphere would show the strong subsidence

Table 3-12 Parameters used to monitor the ENSO cycle

1	2	3	4	5	6	7	8	9	10
Mo/Yr	SST	SOI	850-hPa WI WP	850-hPa WI CP	850-hPa WI EP	200-hPa WI CP-EP	OLR DL	Mo rain	Mo avg rain
May/99	−0.7	+0.1	1.8	0.8	−0.4	0.6	0.9	157	390
Jun/99	−0.9	−0.1	1.0	0.3	−1.0	−0.3	0.7	128	164
Jul/99	−0.7	+0.5	1.1	1.0	−0.1	1.2	0.6	159	142
Aug/99	−1.1	+0.1	1.3	0.5	−0.6	−0.5	1.0	335	159
Sep/99	−0.9	−0.1	1.2	−0.2	−0.9	−1.3	1.4	307	184
Oct/99	−0.9	+0.9	2.3	0.8	−0.1	0.7	1.3	329	234
Nov/99	−1.3	+1.1	1.7	1.4	0.7	1.4	1.4	461	264
Dec/99	−1.6	+1.5	2.8	1.5	0.6	2.1	2.6	540	263
Jan/00	−1.8	+0.7	1.4	1.6	1.1	1.4	2.3	247	315
Feb/00	−1.5	+1.6	3.1	2.1	−0.2	1.6	2.6	160	288

Columns 2 and 4–8 are values between 5N and 5S. SST is the Niño3.4 sea surface temperature (170W-120W); Wind index represents deviation from climatological monthly mean; WP = 135E-180, CP = 175W-140W, EP = 135W-120W, CP-EP = 165W-110W. For 850 hPa wind indices, positive (negative) values imply easterly (westerly) anomalies; for 200 hPa wind indices, positive (negative) values imply westerly (easterly) anomalies. OLR is the outgoing longwave radiation (positive values indicate less cloud) and DL = 165E-165W. Columns 9 and 10 are the actual monthly and the mean monthly rainfall respectively for Fiji. Shaded months indicate peak La Niña conditions (columns 2–8 from the *Climate Diagnostics Bulletin*; rainfall from the Fiji Met Service).

(downward motion vectors) associated with La Niña. The effects of the subsidence produce a drying down to the surface. The monsoon trough, which constitutes the east-west oriented portion of the South Pacific Convergence Zone (SPCZ), is weakened by the strong easterlies, and is also pushed westward. Thus, tropical cyclone development is generally west of the date line.

The north-south oriented portion of the SPCZ becomes anchored between Samoa and the date line. If the SPCZ meanders over Samoa, then the islands are wet. If it meanders away from Samoa, the islands are dry. The Solomon Islands and Vanuatu are also generally drier, but they do experience occasional periods of heavy rains when the SPCZ pushes northward or the monsoon trough surges eastward. During La Niña, these episodic surges are more infrequent and short-lived than normal.

During the Southern Hemisphere dry season, mid-latitude westerlies push equatorward in the vicinity of the date line, bringing heavy rains to a wide band that includes Fiji, Tonga, and New Caledonia. The excessive dry-season rainfall at Suva (Fiji) during La Niña is shown in Table 3-12. Conditions to the east and north become drier as the divergent trade winds strengthen, bringing dry weather to Samoa, Tokelau, Tuvalu, the northern Cook Islands, and Kiribati.

REFERENCES

Allan, R., J. Lindesay, and D. Parker, 1996: *El Niño Southern Oscillation and Climatic Variability*. Mordialloc, Victoria, Australia: Climate Impact Group, CSIRO Division of Atmospheric Research.

Trenberth, K.E., 1997: The definition of El Niño. *Bulletin of the American Meteorological Society*, 78, 2771–7.

Tropical tunas and the ENSO cycle

Gary Sharp

"Normal" conditions in the tropical ocean would be a more or less uniform distribution of tropical ocean surface temperatures that would be warmest at the equator, and dwindle as one moves poleward. The turning of the earth and the location of land masses have made this simplified model of the world take on new forcing. The more usual case is that the eastern portions of equatorial oceans exhibit a thinner warm upper layer during most seasons. In the west there is a concomitant thickening, as warm surface waters are evaporated off the eastern tropical oceans by surface winds of the Walker Circulation and delivered downstream in the form of moisture and heat. The Indian Ocean is the mirror image of this scenario, with the consequences described below.

This thickens the western ocean tropical habitat, where tropical tunas thrive, and heats the upper levels of the ocean to maximum levels. The several species of tropical tunas that inhabit these warm layers are affected in different ways, depending upon the patterns that they encounter at different stages in their development. For example, skipjack tuna (*Katsuwonus pelamis*) are tied to the highly oxygenated upper layers due to their extreme metabolic rates associated with maintaining hydrostatic equilibrium, i.e., so that they do not sink out of their habitat. This situation causes a generic physiological response in that such organisms, be they fish or other, tend to expend more of their available food energies on swimming than on growth. In doing this they tend to be relatively smaller at age, and then tend to develop reproductive capabilities earlier rather than building bigger bodies.

Another group of tropical species tends to drop down into the water column, where, typically, the ocean is cooler, hence less stressful. The limitation in this case is that there must be adequate oxygen available for them to swim, maintain hydrostatic equilibrium, and grow, or they too, will be forced into reproduction rather than somatic growth. The tropical ocean has variously been populated and adapted to by several species of tunas, with unique swimming energetics, hence oxygen requirements, and temperature tolerances, that reflect these tradeoffs. If one were to drop a line with an array of hooks into the depths of the western tropical oceans, as many as six species of tunas can be caught at any one location, but they will have very different distributions amongst the hooks by depth.

Now, if we quickly travel to the eastern tropical oceans, the usual upper ocean upwelling tends to bring oxygen-poor cool water near the surface. This happens for the simple reason that the cooler source waters underlie some of the more productive tropical surface waters on earth, and due to the excess production, the night time metabolism of the plankton tends to extract most of the available oxygen. This reduces the availability of oxygen rapidly with depth, and excludes most of the species of tunas found in the oxygenated western tropical oceans. This means that the same fishing experiment (dropping a line with an array of hooks) within the eastern tropical ocean will yield fewer species. This is observed everywhere except in the eastern Indian Ocean, where everything is complicated by the merger of two major ocean bodies, filtered through a shallow sea environment.

Meanwhile, as the reversal of the Walker Circulation surface winds occurs, the depth distributions of upper ocean tropical water evens out between the two extremes. The primary production patterns reverse, tending to allow more species to invade the eastern oceans, and for those species usually constrained to depths that make them unavailable to surface fishing gear to spend more time near the surface. The western Pacific seine fisheries start to observe more large yellowfin in their catches, as soon as the signature Kelvin wave is released, and the thermocline structure emerges.

To the east, the deepening of the thermocline and the suppression of primary production allow more species access, and deepens the habitat, providing more living space, and making the usually vulnerable tunas less visible, hence less vulnerable to surface fishing techniques. Catch rates tend to decline rapidly, as the signature Kelvin wave arrives, and in situ heating occurs because of night time cloudiness.

The deeper fishing longline tuna gear exhibits generally opposite trends to those experienced by the surface seine gear and must change distribution to enable their gear to be effective – locating deeper, oxygen rich water becomes their objective.

In extreme situations of ENSO warming the expansion of tropical tuna habitat poleward is notable, and makes available to usually temperate ocean fisheries some portion of the tropical tunas and associated species such as marlins, dolphin fish (*Coryphaena spp.*), and some jacks (*Seriola* and others). These invasions by tropical predators can cause havoc in local fisheries where small pelagics and other coastal species can be preyed upon by these voracious predator species. In cold periods, these same coastal regions tend to develop stronger upwelling, hence more local production, and abundances of secondary predator species from anchovies to jellyfishes can bloom. Beyond the usual local predator populations, toward the poles, and seaward, there are entire arrays of usually excluded predators. Offshore of the eastern boundary currents (along the western coasts of continents) lie the more tropical and subtropical elements of the oceanic environment. During local warming, and warming associated with transport of the ENSO tropical ocean heat resources, these predator fields are often overlain onto the coastal environment, suppressing upwelling production, and causing the local fauna to submerge. By making vertical dives, the invaders from offshore can devastate local coastal resources.

On the other hand, during extreme cool events, coastal upwelling can become more intense as well as more frequent, tending to transport more of the production offshore, leaving little behind for emergent fish larvae, or birds, and can in this fashion also devastate local production. It is, in fact, the balance between extremes, the transitional processes, and the dynamic interactions between the various ecological components and compartments that is important, and links these systems into an interactive, sustainable whole. The worst things that happen during any one phase or stage of the ENSO cycle (warm, cold, or average) are good news for another set of participants. The physical dynamic of the ENSO cycle from one extreme to the other is the *reset* button in the patterns of ecological winners and losers.

Part 4

La Niña and the media

The role of the media in ENSO reporting

Michael H. Glantz

Media coverage of the 1997–98 El Niño has been reviewed and assessed by various government agencies, researchers, media specialists and by the media themselves in several countries. One of the major concerns about the coverage has been centered on the "hype" aspects of reports, especially El Niño headlines. Eye catching headlines were used to draw attention to specific newspaper and magazine articles, and sound bites on TV were used to show in a dramatic way adverse weather impacts on people. Scientists, too, have been prone to exaggerate or speculate about El Niño's developments. For example, El Niño had been described by one research organization on various occasions in the month of November 1997 as growing, decaying, and growing again within a span of a few weeks. These rapid changes in forecasts of the fate of what was the 1997–98 El Niño were received in Peru by researchers with access to the Internet. Needless to say, conflicting comments from the same person at a reputable US government facility caused considerable confusion. Sensational, often misleading, headlines created a "doom and gloom" perception of El Niño. It got to a point where newscasters had to report something – anything – that might, even remotely, be linked to El Niño. El Niño's reputation did not stand a chance; it was blamed for just about everything that took place while it was in progress, and even well beyond its demise.

La Niña Summit panelists pointed out that the media do not speak with one voice nor do they operate individually according to the same ethical standard. As a result, each story appearing in the media may end up having been treated by a different set of rules. The result is that the media are varied in the way they report on El Niño and La Niña and on their environmental and societal impacts. Objective, scientifically based reports are not necessarily what the media produce.

Editors as well differ in the way they treat the articles of their scientific writers. While some seek to present the ENSO situation so as to include the uncertainties of the science or of the impacts, others tend to rely more on the sensational and speculative aspects, presenting stories in which scientific facts give way to eye-catching adjectives.

At the La Niña Summit, a participant mentioned that "the press covers the news, not necessarily only good things." Thus, there had been a tendency to report El Niño (and now La Niña) in negative terms, e.g., to report about adverse aspects of El Niño and La Niña. This can be seen from the headlines of various El Niño articles and magazines (Figure 4-1).

Another participant noted that the headlines used to introduce El Niño stories in the printed media tended to track quite well the changes in the attitude of the press toward the ENSO warm event/cold event cycle, as it

El Niño drought could affect 25 million Africans   El Niño affecting coming seasons   Bond Forecast Is Clouded By El Niño

El Nino reduces energy demands

Según congresista peruano
El Niño impedirá conflicto   '$1 BILLION RAINS'   Another soaking for El Nino

Outguessing El Niño
Commodities markets feel the tension   Fujimori against El Niño

El Niño será catastrófico   El Niño mantiene activos incendios en Indonesia

Inundaciones por El Niño en Brasil y Uruguay   Weather experts fear the worst is coming.   El Niño bolsters Hurricane Linda

El Niño upheaval

Рынки под влиянием эффекта «Эль Нино»

Wily El Niño has forecasters scratching head

El Niño looms, but it won't be all bad

El Niño could be easy on winter heating bills   Girding for El Nino
For some, 'It's almost reached the level of panic'

El Niño: Preparing for the Worst

Devastating El Niño Forecast
Scientists fear ecosystem chaos   El Nino effect warms waters off California   EL NINO PEAKS -but will second wave come?

Fig. 4-1 El Niño headlines taken from various newspapers around the world.

developed in 1997 to 1998, from the forecast of the onset of an El Niño to its demise and to the possible onset of La Niña.

It was suggested that, in order to capture the broadest audience, the media tend to present scientific information that is geared to the 5th to 7th grade levels (in the US education system). This approach may sell papers, but it can lead to misinformation and misunderstanding about ENSO warm and cold extremes in attempts to keep the science rather simple and the story readable, if not sensational.

Several of the following problems associated with reporting on La Niña and El Niño have been identified. A serious communication problem exists between the media and the scientific community, because of several factors: very few science writers have a scientific background; there is a tendency of both scientists and the media to overstate (the media tend to highlight extremes and shifts, whereas the scientists tend to present their research efforts in a more positive light); they acknowledged that the media have a penchant for bad news and so there is a tendency for them to highlight adversities rather than to present scientific information they believe will not capture the public's interest; time factors also put pressure on the media to produce a story early, if not first. Time can also become a problem in the sense that it is difficult for the media to sustain interest in a scientifically based issue; the science reporters and writers are put in the position of writing about a natural process (i.e., El Niño or La Niña) that is occurring over a relatively long period of time (12–18 months), a phenomenon that scientists have different views on how to describe or to quantify.

According to the media representatives, the scientific community often passes the responsibility of describing this intricate phenomenon on to the media. The media often have problems with providing a balance of views in their articles, and with regard to the sorting out of expertise: which "experts" should they trust? One problem has been the apparent need for the media to report minority views that are not championed by the scientific community. Yet, such minority opinions are often given equal space or time as the dominant scientific view.

Panel members contended that it was really the sustained public interest in the possible impacts of ENSO's extremes that led to the continuous coverage of their forecasts and development and expected impacts. This blitz-like coverage led to charges against the media that they were hyping El Niño, and later, La Niña. A participant suggested that for all practical purposes "El Niño might as well have been the Spanish word for hype." Several members of the media panel felt that very few scientists were really good at dealing with the press. They also contended that a large part of the scientific community looked upon the press in a condescending way.

Another panelist suggested that scientific researchers should not view reporters as friends. The reporter has a mission to produce an "attractive" news report and to meet a deadline, one that has most likely been set by someone else. If the scientist has a point to make it is up to her or him to get that point across clearly. For the most part, reporters try to fact-check the contents of their stories but, given the strict deadlines imposed on them, foolproof fact-checking is often not possible.

Once they appear, sensational headlines are quoted repeatedly. They are designed to capture the reader's attention even though it may have little to do with the text or tone of the ENSO-related article. Reporters, however, do not choose their headlines; editors do.

Time is a very important factor when producing an article on ENSO. Often, the reporter does not have time to hold a second interview with the scientist(s) that s/he interviewed at first. As a result of the pressure of time, there is a tendency for a reporter to go back at first to the familiar sources that s/he trusts, even if that person is known not to be the best source of that particular piece of information.

For some media, the reporter/writer is supposed to see and discuss any changes made by others to their article. However, this is not always possible and, as one participant suggested, "gremlins make changes in their stories," that is, errors have been inadvertently introduced into the final version by others.

A common complaint made by scientists about interviews with the media is that they had been misquoted or that their comments had been taken out of context. Media representatives responded to this point by noting that their stories are tailored to meet a space or a time slot constraint. They asserted that, while they do not change a quote, they may have to shorten it.

To generate a more positive view from the public about forecasts of El Niño, La Niña, or their impacts, it is necessary for the media to report them in probablistic terms. However, the media, among others (including scientists), believe that the public has great difficulty in understanding probabilities. One summit participant noted that he had been made to look foolish by the media, because he spoke in terms of probabilities of occurrence of precipitation in southern California. When the rains did not occur when expected, the media referred to El Niño as "El No-Show." Yet, the rains did come and were as heavy as predicted.

The media panel noted that scientists need to work more closely with the media so they can identify ways in which they could educate one another. Workshops organized to center on the interactions between these groups would be a fruitful way to improve the understanding of the strengths and weaknesses in reporting about ENSO's extremes and their impacts to the general public.

One media specialist noted that it was necessary to scrutinize their sources more carefully, given the proliferation of websites on the Internet and with no one in charge to oversee the reliability of their content. Trying to determine which sites were providing reliable information was a concern to some media representatives. The same care must be taken in order to sort out biases that accompany supposedly objective scientific information. For example, were the forecasts of the onset of El Niño (or of La Niña) made as early or as successfully as some researchers have suggested? Were they as reliable as suggested? A recent article in *Science* (Kerr, 1998) suggested that "Models win big in forecasting El Niño" with respect to an advanced warning of the onset of the 1997–98 El Niño. However, such a conclusion has not been supported by retrospective assessments of the forecasts (e.g., Barnston et al., 1999). Suggestions about devising a way to rate the reliability of various websites generated a heated debate about the fear of censorship of websites that presented perspectives opposed to mainstream views.

There are real legitimate differences among scientists about scientific perspectives and opinions about the ENSO cycle. This is a troublesome situation for the media, which would prefer to report such stories in black-and-white terms.

There was confusion in media reporting between forecasting the onset of the El Niño phenomenon itself and the forecasting of its environmental and societal impacts several months after its onset. Such confusion is generated, for the most part, by the scientific community. Claims about El Niño forecast successes sometimes related implicitly to the forecast of its impacts, even though the forecast of the El Niño phenomenon was issued about the same time that SST observations showed that an El Niño had already begun to develop. These, however, are not the same. Successes for forecasting impacts do not translate into successes for forecasting El Niño.

It is clear that the 1997–98 El Niño became a global story that was carried in the media worldwide. Comments by several foreign participants at the summit were made about El Niño coverage in their national media (e.g., Australia, Ecuador, Ethiopia). For example, the Australian situation was confusing to the public because, in addition to its nationally developed El Niño forecasts, the media in Australia were bombarded with forecasts (often conflicting) from various US agencies (see Kestin, this volume). It seems that few places matched the amount of media coverage in the United States and in Australia. Media interest in La Niña followed on the coat tails of interest in this major 1997–98 El Niño. It is clear that there is less known about it, as suggested by the newspaper headlines devoted to the then-emerging La Niña (see Figure 1-7). Coverage of La Niña over the following two years had been considerably less than it had been for an El Niño of comparable length.

Key points

1. The media do not operate as a homogeneous unit that speaks with one voice. They do not operate by one set of rules or share the same reporting ethics. These vary from media to media, within a specific media, and among reporters and editors.
2. Media are attracted to stories that are unusual or have adverse consequences. They do not necessarily focus on good news or on protracted processes.
3. They are responding to the news interests of the general public.
4. Presenting La Niña and El Niño information in probabilistic terms is the correct way to go. However, participants in general felt that the public was not capable of understanding or using such information. Others; however, believed that it was time to prepare the public for such probabilistic statements.

Note

Participants on the Media Panel included Bill Meck (WTHR TV, Indianapolis, Indiana), Gary Robbins (Orange County Register, California), Joe Virengia (Associated Press, Denver Bureau), Madeline Nash (*Time Magazine*), Mary Miller (Exploratorium, San Francisco), and John Kermond (Office of Global Programs, NOAA, Washington, DC).

REFERENCES

Barnston, A.G., M.H. Glantz, and Y. He, 1999: Predictive skill of statistical and dynamical climate models in SST forecasts during the 1997–98 El Niño episode and the 1998 La Niña onset. *Bulletin of the American Meteorological Society*, 80(2), 217–43.

Kerr, R.A., 1998: Models win big in forecasting El Niño. *Science*, 280, 522–3.

Forecasting ENSO in Australia: The importance of good communication

Tahl Kestin*

The forecast of the 1997–98 El Niño was not a forecasting success in Australia. Many people misunderstood ENSO-based rainfall forecasts, and therefore took inappropriate actions for dealing with El Niño. The misunderstanding was mainly the result of difficulties in communicating the forecasts, and involved the Australian Bureau of Meteorology, the media, and the public. These difficulties are common to situations where forecasts involve substantial uncertainty – such as forecasting El Niño and La Niña. Thus, Australia's experience with the 1997–98 El Niño highlights the importance of communication in making ENSO forecasts useful to society.

ENSO explains about 30 percent of Australian rainfall variability. During El Niño events rainfall tends to be below average, while during La Niña events rainfall tends to be above average. These variations in rainfall have an observable effect on agricultural yields in Australia, so rainfall forecasts based on ENSO *should* be useful to farmers. The Australian Bureau of Meteorology has been issuing ENSO-based seasonal rainfall forecasts for more than 10 years. However, despite the knowledge and experience gained by the Bureau over the years, few people realized the potential benefits of the forecasts during the 1997–98 El Niño.

In May 1997 the Bureau, noting the developing El Niño, began issuing predictions of a higher likelihood of below average rainfall in many parts of Australia. As it happened, rainfall during the 1997–98 El Niño was generally below average, yet the Bureau was accused of getting the rainfall forecasts wrong and, therefore, misleading farmers. These accusations stemmed from the public's *expectation* that the El Niño would cause a devastating drought, whereas unanticipated but timely rainfall in the spring – the highest recorded for any El Niño – saved many crops.

The public's expectation that El Niño would cause a devastating drought was inappropriate because the effects of ENSO on rainfall in Australia are, to a large extent, uncertain. The ENSO-rainfall relationship is clear over several years, or over large spatial scales. However, the areas affected, and the severity of the effects, vary from one event to another. This uncertainty is reflected in rainfall forecasts through the use of probabilities. An El Niño increases the probability of below average rainfall, but average or above average rainfall is still possible. So, why did the public expect a severe drought, given the uncertainty contained in the rainfall forecasts? To a large extent, the public's expectations were the result of a breakdown in the communication process between the Bureau, the media, and the public (Figure 4-2).

Although the Bureau issued probabilistic forecasts, there was a gap between what the Bureau thought they were saying in the forecasts and how the public understood it. For example, a significant likelihood of below average rainfall was interpreted as no rainfall (i.e., drought) (Nicholls and Kestin, 1998). This interpretation was probably the result of the language the Bureau used to express probabilities (Kestin, 2000), and people's natural tendency to prefer categorical forecasts (Nicholls, 1999).

The public's expectations were shaped further by media hype surrounding El Niño. When the media reported on El Niño, they tended to report the forecasts as categorical, rather than probabilistic, statements. They also emphasized the link between El Niño and severe drought.

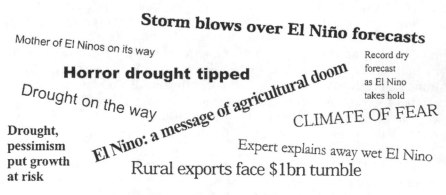

Fig. 4-2 Australian headlines about El Niño.

These characteristics of the El Niño media reports reflect principles on which the media operate: bad news sells papers, whereas statements of uncertainty do not (Kestin, 2000). In addition, several international meteorological agencies were reported in the Australian media as saying that Australia can expect a devastating drought.

The actions that farmers took in response to the forecasts clearly showed that they misunderstood the forecasts and El Niño in general. There were many stories of inappropriate actions taken by farmers – for example, panic-driven asset selling, not planting crops, following the daily progress of El Niño, and disregarding the forecasts "in hope that they will be wrong." It also appears that many farmers relied only on the information supplied in the media, and had little prior knowledge of the nature of El Niño.

Lack of knowledge on how to interpret and use ENSO forecasts caused many farmers to expect El Niño to have particular effects, and therefore to take inappropriate actions to cope with them. These inappropriate expectations led to the forecasts, in some cases, leading to more harm than good. They also caused the Bureau to appear unreliable, because when good Southern Hemisphere spring rains came in 1997, they were not seen as a "chance" event, but as proof that the Bureau's forecast of drought was wrong.

La Niña presents the same difficulties with probabilistic forecasts as does El Niño. The main difference between the two – at least in Australia – is that El Niño is well known and generally feared, whereas little notice has been taken to date of La Niña. La Niña's modest reputation may make it harder to get the public and the media interested in it – a problem from which El Niño no longer suffers.

There were several reasons why forecasts during the 1997–98 El Niño were not used to their full potential in Australia. These reasons are linked to difficulties – experienced by the forecasters, the media, and the public – with communicating the forecasts. These problems are neither new, nor are they confined to ENSO forecasts. And there are ways of overcoming them, though these ways sometimes require considerable effort (for example, see Glantz, 1995; Kestin, 2000). However, we first need to recognize the importance of communication; without good communication, ENSO forecasts will not be as useful as they could be. Despite all the difficulties, we believe that improved focus on communication strategies should become a priority for forecasters, users, and the media.

Note

* Tahl Kestin at the time of publication has a postdoctoral position at the IRI, PO Box 1000, Palisades, NY 10964, tahl@iri.ldeo.columbia.edu.

REFERENCES

Glantz, M., 1995: Executive summary of workshop discussions. *Proceedings of the Usable Science Workshop II: The potential use and misuse of El Niño information in North America.* Boulder, CO: Environmental and Societal Impacts Group, National Center for Atmospheric Research.

Kestin, T., 2000: *Variations of Australian Climate and Extremes.* Ph.D. Thesis. Clayton, Victoria 3800, Australia: Monash University.

Nicholls, N., 1999: Cognitive illusions, heuristics, and climate prediction. *Bulletin of the American Meteorological Society*, 80, 1385–98.

Nicholls, N. and T. Kestin, 1998: Communicating climate. *Climatic Change*, 40, 417–20.

Awareness of ENSO events in Japan

Mikiyasu Nakayama

Awareness of ENSO events dramatically mushroomed during the "historical" 1997–98 El Niño period. Many people in "weather-dependent" sectors of society became worried about El Niño's possible adverse effects on their own sector. For example, the CEO of a large chain of retail stores in Japan expressed his concern in July 1997 about the possibility of reduced sales of summer clothes. He did so because the mass media warned of the possibility of a "cold summer" and a "warm winter" in the Northern Hemisphere, as the major effects of El Niño. When the emergence of an El Niño had been confirmed by the Japanese Meteorological Agency in early June 1997, numerous articles about the possible adverse effects of El Niño (a cold summer in particular) were featured by Japanese newspapers in June and July 1997. The sale of beer and soft drinks during summer was assumed to be reduced. Reduction in tourism for the summer vacation period was also a matter of concern for the travel industry. It was suggested that the Nagano Olympic Games, planned for early 1998, would suffer from a paucity of snowfall because of a warm winter. The mass media also hinted that the decrease by 1 °C of average temperatures in July and August should lead to a decrease in consumption in general by 0.3 percent. Moreover, agricultural production was expected to decrease by 7.3 percent, if the summer in 1997 turned out to be as cold as had been in 1993, which was then considered the worst cold summer in the past decade or two. A famous Japanese think-tank went so far as to predict that the GDP of

Japan in 1997 would shrink by 0.2 percent because of El Niño (*Saga Shimbun*, 1997).

Views varied about the extent of El Niño's effect on temperature in summer, as cited in newspapers. One article mentioned that a cold summer was recorded in seven summer seasons out of the nine El Niño events that had occurred since 1972, while another article mentioned that the possibility of having a cold summer during an El Niño event was estimated at no more than 40 percent. It was apparent that there is no solid consensus, as yet, among researchers about the effects of El Niño on Japanese weather in summer.

The summer of 1997 turned out, in fact, to have been a warm summer rather than a cold one. Agricultural production for that year, in particular paddy rice, was above normal. Most of the "predictions" that appeared in the newspapers proved wrong, namely the emergence of a cold summer and its related impacts. An exception was the number of typhoons that emerged in 1997, in that many more typhoons were observed in the Pacific than in normal years, as predicted by many meteorologists.

Public awareness about El Niño skyrocketed during the onset of the 1997–98 El Niño event, while awareness about La Niña was still quite low, even in 1999, which was the middle period of a lengthy multiyear La Niña. Table 4-1 shows the number of articles on ENSO extreme events that appeared in *Saga Shimbun*, a local newspaper published in southern Japan. Most of the ENSO-related articles that appeared were actually developed by a news agency specializing in news abroad and delivered to local newspapers. It is, thus, safe to assume that the tendency for reporting about ENSO events should be the same for other Japanese newspapers, both local and national.

Major national and local newspapers featured an article "Sign of La Niña" by the news agency *Saga Shimbun*, about the emergence of La Niña on its front page in early July, with an imagery of sea surface tem-

Table 4-1 Number of articles in *Saga Shimbun* on ENSO events

Year	On El Niño	On La Niña
1994	9	1
1995	5	0
1996	2	1
1997	46	2
1998	42	7
1999	11	6

El Niño occurred from March 1997–May 1998. La Niña began in mid-1998 and continued into the year 2000. However, this survey was carried out in mid-1999, so that not all La Niña articles were counted in that year.

perature of the Pacific taken by the Japanese TRMM satellite and released by NASDA, the Japanese space agency (*Saga Shimbun*, 1998). A few articles on La Niña appeared thereafter in other newspapers as well, although the number of such articles was still not as numerous as those that appeared during the 1997–98 El Niño event.

The following possible impacts of La Niña were mentioned in newspaper articles: (1) a cold winter in Japan and a subsequent increase in fuel consumption; (2) fewer typhoons would hit in Japan, which could lead to less damage to houses, roads, railways, etc.; (3) a hot summer, which may induce more electricity consumption for air conditioning; and (4) a shorter rainy season, which leads to shortages of water for paddy fields in the summer. These effects are clearly based on the belief that La Niña is the opposite of El Niño. Few articles were accompanied by statistical figures.

The awareness of La Niña in Japan is still low, especially in comparison to El Niño after mid-1997. The "possible effects" of La Niña mentioned in the mass media were mostly based on an assumption that El Niño and La Niña are exactly opposite phenomena, and were not necessarily based on the solid objective output of scientific research on the La Niña phenomenon. It is important for the research community in Japan to identify the alleged "symmetry" aspects with regard to the societal and environmental impacts of ENSO warm and cold extremes.

REFERENCES

Saga Shimbun, 1997: Fuji-Souken estimates impacts of El Niño. *Saga Shimbun*, 20 July.
Saga Shimbun, 1998: Sign of La Niña. *Saga Shimbun*, 7 July.

Science and the media

John L. Kermond

The following remarks are made from the perspective of someone who has interfaced between the science involved in ENSO research and the media that report on it, in both print and television. The remarks are in no order of importance, but they do reflect some operating principles that I have found to be appreciated by the media and that have been successful for the federal agency or program in getting the requisite message both "out" to the public and comprehended.

1. Make a policy of communicating what we know, what we do not know, and the significance of both. For instance, with respect to climate change, there is much we have learned and much we have yet to fully understand; like the role of clouds or the inadequacies of present-day computer models of complex integrated systems. But we are using the best minds, with the most advanced tools available, and we are making giant strides in understanding, monitoring, and predicting the earth system's behavior. The increasing ability to reliably forecast an ENSO event or some of its teleconnections is an excellent example of this progress.

2. Provide the reporters or the camera crew with everything that will make their story formulation easier or make their "sell" to the news editor easier! In September 1997, NOAA/OGP provided 30 minutes of "El Niño" (B-roll [background footage] on Betacam SP) tape to attendees at the Radio Television News Directors Association (RTNDA) annual meeting in New Orleans, Louisiana. From direct feedback, we

learned that this tape was instrumental in reporters doing stories on the then-predicted El Niño event, and gave them additional footage to use in subsequent stories as the impacts of this very strong event unfolded. (Incidentally, in my opinion, the US media, in general, did an excellent job in communicating and educating the global public on this ocean-atmosphere phenomenon. In fact, WFAA, an ABC-TV affiliate in Dallas, Texas, recorded its largest-ever market share (e.g., the number of viewers) for the third part of a special show on El Niño which it aired during the important US ratings week against the very popular NBC show "ER" – a TV series shown nationwide during prime time.

3. We need to be cognizant of modern news formats. For example, we are now very used to 7- to 12-second long changes of images on television. (NB: You can test this yourself from outside looking in on the light changes from a TV set operating in a darkened room.) This is good for sound bites. The bottom line for spokespersons or interviewees is to have their sound bites formulated and ready and to express them with vigor, enthusiasm, and obvious sincerity. Cameramen and editors can hear a sound bite from far away! This is what is likely to get on the TV news – where most stories run only for an average of 100 seconds!

4. The print media is pitched at an equivalent 8th-grade (US) educational level of reading comprehension. This demands that scientists, in particular, steer their comments away from scientific jargon and from bureaucratic acronyms. Instead they should use common, understandable words. For example, use the word "mountains" instead of "orographic"; use "space" and "time" and get rid of "spatial" and "temporal." An old rule of thumb is that if the word is a difficult one to spell, then it should be avoided verbally!

5. Another fact of life in the media business is "if a news item bleeds, it leads." Many editors favor a "man-bites-dog" story, i.e., the unexpected and unusual event. Most editors will ask the fundamental question – "What's the news?" This can be most disconcerting for many scientists who want to report with much enthusiasm that "a glacier has moved a millimeter." The main point here is to relate the science to the person in the street: What does it mean? How does it affect or effect? What are the impacts, both beneficial and adverse? Rely on pointed analogues to enhance the public's understanding of complex scientific processes. For example, the chemistry of the atmosphere has been described by a leading authority as akin to a Cajun soup – nitrogen "base" with lots of different "spices," and rarely the same at any given moment in time or space.

6. Another "given" in the media industry is the need to show some

"balance" in a story. It is reasonably easy to find someone in the scientific community with a different point of view or a bias. The real tricky part for the reporter is when she or he has to report on a statement that has been endorsed by more than 2,000 scientists from around the globe (e.g., the Intergovernmental Panel on Climate Change, or IPCC), but has been opposed by a relatively few vocal individuals (e.g., the global warming issue). To my mind, this is not a way to achieve balance, where scientifically none may be warranted. It makes the task of the science writer harder. The bottom line here is as follows: it is necessary to think through what the known antagonists might counter with and develop very succinct responses which can be used in a question-and-answer session, or can be referred to frequently throughout a briefing. This ties in directly with my first point. By way of example, the promoters of the experiment that was to track the passage of sound throughout the world's oceans should have anticipated that opposition would come from those concerned for the safety of certain marine mammals. They obviously did not, and the experiment was initially delayed and apparently abandoned after only one attempt (Hartz and Chappell, 1997).

7. Government agencies have a negative tendency to want to control briefings, to control access to scientists, and to control their messages. Obviously, some sense of decorum must exist, but, from my experience, many journalists are "put off" by a government representative announcing the rules of a briefing. It is far better for that representative to suggest ways to make the briefing go more smoothly, or faster. This uncovers another negative tendency of both government agencies and scientists themselves: the tendency to be long-winded. I once experienced a US Congressional briefing to a mixed audience of media and congressional staff that was slated for two hours. The time was theoretically cut in half after some swift negotiation. The CNN cameraman told me afterwards that he used more than an hour of tape but that he never once heard a sound bite. I set him up for a one-on-one interview and this was the piece that went to air in reporting the story. All of the other presenters had missed their chance to provide useful information to decision makers and the public.

REFERENCE

Hartz, J. and R. Chappell, 1997: *Worlds Apart: How the Distance Between Science and Journalism Threatens America's Future.* Nashville, TN: First Amendment Center.

Part 5

More thoughts on La Niña

La Niña and an evaluation of current ENSO forecast skill

*Anthony G. Barnston**

Definition

La Niña can be defined in terms of the rank of the average sea surface temperature (SST) anomalies in a designated region in the tropical Pacific, such as Niño3.4 (Barnston et al., 1997), or farther west near 140°W–170°E. If the SST ranks in the bottom portion of the distribution for the given season, then a La Niña state can be said to exist. The limits of that bottom portion are arbitrary: it may be the bottom 20 percent, the bottom 15 percent, or some other limit. When the limit is too strict, the events included are strong, but the sample size is undesirably small. When the limit is more lenient, a larger sample can be included, which is good for reducing the influence of atypical cases, but it runs the risk of including La Niña events of very different strengths and "flavors" (see Trenberth, this volume).

The above ideas apply also to additional indicators of the ENSO state, such as low level winds, the sea level pressure field, or subsurface sea temperature. Rather than trying to form categorical boundaries for the ENSO state, it might be preferable for most purposes to measure it on a continuum using a numerical rating as is done in the Multivariate ENSO Index (MEI) or the Niño3.4 standardized SST anomaly. Then, users can define their own categories if desired, based on their particular application. (See Figure 2-11 for a map depicting the five Niño regions.)

Teleconnections

Using a specific definition of La Niña, teleconnection patterns locally and globally can be identified. As the length of the period of high quality climate data expands (both because time keeps passing and because of better ways to reconstruct the data of the past), more precise knowledge of the teleconnections will become possible. Additionally, improved dynamical models may offer refined teleconnection expectations, even without the proof of empirical data. A complicating factor, however, may be climate change. If the character of a "typical" ENSO episode gradually changes, as a result of global warming or for other natural reasons, it stands to reason that its teleconnections will also be subject to change.

Climate change

Over the last 30 years or so, there has been a gradual tilt toward an increase in the number of El Niño events at the expense of La Niña. This is because a global climate change pattern is partially correlated spatially with the ENSO pattern (in the tropical Pacific SST). In recent times, therefore, it has been less common to have a strong La Niña. The global climate change in the tropical Pacific has consisted of an increase in the SSTs near the dateline, with a smaller increase in the eastern portion of the Pacific basin. This pattern bears some resemblance to the El Niño-like conditions that dominated during the early 1990s (Zhang et al., 1997).

Symmetry

El Niño and La Niña are somewhat – perhaps largely – mirror images of one another. But this symmetry only holds up to a certain point, beyond which differences appear. These differences are based on the asymmetric physics of enhanced convection versus suppressed convection; once convection drops to near zero, additional cooling of the SSTs no longer has an effect on convection. El Niño does not behave that way: the stronger the event, the more the anomalous convection and the consequent changes to the climate in remote locations. The greatest positive anomalies of convection in El Niño are located in the 120°–180°W region, while the greatest negative anomalies of convection in La Niña are positioned farther west, from 150°W to 160°E (i.e., the Niño4 region). The asymmetry of El Niño and La Niña is discussed in Hoerling et al. (2001) (see also Zebiak, this volume).

Attribution

Attribution of distant climate anomalies to the occurrence of an El Niño or a La Niña is a very tricky issue, because a lot of non-ENSO processes occur during ENSO extreme events. Certain overly eager scientists and the media used their power of embellishment in ascribing certain climate-related impacts to the 1997–98 El Niño, such as the October 1997 snowstorm in Colorado, and, to a lesser extent (since it was partly confirmed by a dynamical model), the January 1998 ice storm in Canada and New England. The possible influence of El Niño on individual weather events is discussed in Barsugli et al. (1999).

The most obvious symptoms of the recent El Niño, about which there is little doubt, are (1) winter storms along the California coast, and (2) lots of winter rainfall in Florida, winter drought in Micronesia and Indonesia, and abnormal winter warmth in the upper US Midwest and south-central Canada. Shorter-term extreme weather events may have been modulated by the El Niño, but probably not caused by it per se. Modulation may imply exacerbation, mollification, or just neutral change. For the 1998–99 and 1999–2000 La Niña, a correct attribution would be the winter warmth observed along the southern tier of US states during both Northern Hemisphere winters.

Forecast skill

Forecasting El Niño is still fraught with problems. And, to date a reliable long-range El Niño forecast system has not been developed. The situation is similar for forecasting the onset of La Niña. The CPC (Climate Prediction Center) in NOAA did very well in the winter of 1997–98 forecasting the climate effects, once their forecasters knew (by summer 1997) with near certainty that the El Niño would exist during the winter of 1997–98. As for forecasting in March 1997 the onset of an event, most of the statistical and dynamical models did forecast some degree of warming, but none predicted the high amplitude of the event. The skill level can, therefore, be called mediocre. Forecasting La Niña is generally of the same order of difficulty as forecasting El Niño. The physics is different (e.g., less convection rather than more, as noted earlier), but representing the physics correctly, in dynamical models, remains a similar problem for both extremes of the ENSO cycle.

Because ENSO extremes usually begin to develop in the boreal (Northern Hemisphere) spring or early summer and persist through the following winter, forecasting impact tendencies in extratropical North America for the winter (when impacts are most pronounced) with five months of lead time is, relatively speaking, not difficult. This only requires good observations of the summer ENSO state in the tropical Pacific, and knowledge of the winter teleconnections. Because of the strength of the 1997–98 El Niño and the consequent skill of the five-month lead forecasts of US impacts in the winter 1997–98, the success of these forecasts was noticed to an unprecedented extent by the media and, hence, the general public. However, forecasting impacts in the austral (Southern Hemisphere) winter which occur simultaneously with the initial appearance of an ENSO extreme (e.g., in Chile, Uruguay, Kiribati, Ecuador, and Peru) requires forecasting the boreal spring/summer onset of ENSO events with several months of lead time. This latter task is difficult, as forecast performance remains mediocre as evidenced by the fact that the formal announcements of a major El Niño did not occur until May 1997, when the El Niño was already becoming strong. This left little time for users in some of the above regions to prepare for its potential impacts. Clearly, continued scientific effort and computer resources are needed to advance our understanding of tropical and global ocean-atmosphere interactions and their simulation in coupled models.

Barnston et al. (1999) reviewed forecasts of ENSO conditions in 1997–98. They reviewed the output of 15 dynamical and statistical models for the onset and maturation of the 1997–98 El Niño and the onset of the development of a La Niña in 1998. Their review noted that, while most of the models had forecast some degree of warming one to two seasons prior to the onset of the El Niño in the boreal (Northern Hemisphere) spring of 1997, none had detected its strength until an El Niño was already in the process of becoming strong in early summer. Neither the dynamical nor the statistical models, as groups, performed significantly better than the other during this episode. The 2–4 best-performing statistical models and 1–2 of the best-performing dynamical models forecast SST anomalies of about $+1\,^{\circ}C$ (versus the 2.5 to $3\,^{\circ}C$ observed) prior to any observed positive anomalies. Their review also noted that the most comprehensive dynamical models performed better than the simple dynamical ones. Once the El Niño had developed in mid-1997, a larger number of models were able to forecast its peak in late 1997 and its dissipation and reversal to cold conditions in late spring and early summer 1998.

Monitoring

The monitoring for La Niña is essentially the same as for El Niño (for a different view see Busalacchi, this volume). Rather than detecting westerly wind bursts in the western or central Pacific, detecting stronger than normal trade winds becomes important. The former may require somewhat more highly resolved measurements in both space and time, since small-scale bursts, sometimes related to the Madden Julian Oscillation (MJO), may end up having major consequences for the onset of El Niño. In both cases, i.e., La Niña and El Niño, measurement of the subsurface sea temperature is very important. This is because sea temperature anomalies below the surface in the western tropical Pacific often tend to become the SST anomalies in the central and eastern Pacific one to two years later (Smith et al., 1995).

Forecaster-user communication

Glantz (1999) reviewed the verbal summaries of ENSO forecasts that had been submitted for inclusion in the Experimental Long-Lead Forecast Bulletin (ELLFB) produced by NCEP/CPC. These bulletins were issued to users worldwide before and during the 1997–98 El Niño event. (NB: After the December 1997 issue, the ELLF bulletins were issued by COLA.) These summaries were found to contain verbal ambiguities, when examined from the users' points of view. Given the need for forecasts to be expressed verbally and also to be precise enough for meaningful use and verification, a simple numerically based verbal classification system for describing ENSO-related forecasts was developed (Barnston et al., 1999).

Key points

1. El Niño and La Niña, as extreme categories, may not be necessary if the ENSO state is measured on a continuum.
2. Teleconnections, or preferred remote responses to ENSO episodes, may gradually change somewhat over the coming decades, as global climate changes cause ENSO to change in its detailed character.
3. The performance in forecasting the onset of the 1997–98 El Niño was largely mediocre.
4. Dynamical models as yet do not outperform the statistical ones, with respect to forecasting El Niño. However, in the future the dynamical models are expected to outperform the statistical models.
5. The communication between forecasters and users still leaves something to be desired. It appears that neither really understands how the other thinks and what the other does or does not understanding about ENSO forecasts and their application to societal needs (see Kestin, this volume).
6. Forecasting the tendencies of teleconnected impacts in North America

for the winter season during the peak of an El Niño event at lead times of about five months is not difficult, relatively speaking, because most El Niño episodes make their initial appearance in the early boreal summer.

Note

* This research was initiated while the author was at NOAA's Climate Prediction Center in Washington, DC.

REFERENCES

Barnston, A.G., M. Chelliah, and S.B. Goldenberg, 1997: Documentation of a highly ENSO-related SST region in the equatorial Pacific. *Atmosphere-Ocean*, 35, 367–83.

Barnston, A.G., M.H. Glantz, and Y. He, 1999: Predictive skill of statistical and dynamical climate models in SST forecasts during the 1997–98 El Niño episode and the 1998 La Niña onset. *Bulletin of the American Meteorological Society*, 80(2), 217–43.

Barsugli, J.J., J.S. Whitaker, A.F. Loughe, P. Sardeshmukh, and Z. Toth, 1999: Effect of the 1997–98 El Niño on large-scale weather events. *Bulletin of the American Meteorological Society*, 80, 1399–411.

Glantz, M.H., 1999: *Reducing the Impact of Environmental Emergencies through Early Warning and Preparedness: The Case of the 1997–98 El Niño-Southern Oscillation*. First meeting of team leaders, 8–10 July, Geneva, Switzerland. Full report on line at: www.esig.ucar.edu/un/geneva.html (hard copy available upon request from ESIG/NCAR, Boulder, CO).

Glantz, M.H., 2001: *Currents of Change*: *El Niño and La Niña Impacts on Climate and Society*, 2nd ed. Cambridge, UK: Cambridge University Press.

Hoerling, M.P., A. Kumar, and T.-Y. Xu, 2001: Robustness of the nonlinear climate response to ENSO's extreme phases. *Journal of Climate*, 14(16), 1277–93.

Smith, T., A.G. Barnston, M. Ji, and M. Chelliah, 1995: The impact of Pacific Ocean subsurface data on operational prediction of tropical Pacific SST at the NCEP. *Weather Forecasting*, 10, 708–14.

Wolter, K. and M.S. Timlin, 1998: Measuring the strength of ENSO events – How does 1997–98 rank? *Weather*, 53, 315–24.

Zhang, Y., J.M. Wallace, and D.S. Battisti, 1997: ENSO-like interdecadal variability: 1900–1993. *Journal of Climate*, 10, 1004–20.

Cold events: Anti-El Niño?

*D.E. Harrison and N.K. Larkin**

Although it has become common to refer to the extremes of the El Niño/ Southern Oscillation (ENSO) phenomenon as El Niño and La Niña, we prefer instead to use the terms "warm event" and "cold event." The following is a brief summary of ongoing research on the common patterns of the ENSO phenomenon, with the focus here on cold events. We first address how to define when a cold event is underway and present the years that are identified as cold event years according to several different criteria. We then continue by comparing a composite of ocean sea surface temperatures (SSTs) for the last nine cold events with a composite of the 10 warm events during the same period (1946–95). The statistically significant anomalies are shown, as computed by removing the climatological seasonal cycle. We find that to a considerable extent warm and cold event years are qualitatively inverse to each other in the equatorial Pacific. However, away from the equator there are noteworthy differences. Finally, we examine the continental US seasonal weather impacts of ENSO in terms of temperature and precipitation anomalies associated with cold event and warm event periods.

The full version of our work is found in a series of papers, both published and in press The warm event period SST results can be found in Harrison and Larkin (1998a), and there is an earlier paper on the sea level pressure (SLP) patterns (Harrison and Larkin, 1996). Another article in *Geophysical Research Letters* discusses the warm event US impacts (Harrison and Larkin, 1998b). The manuscripts detailing the cold event results are in press (Larkin and Harrison, 2002). In these papers, we identify the global ocean surface composite anomaly patterns during ENSO that are statistically significant and those which are also typical of the post-World War II events. Sea surface temperature and pressure, zonal and meridional wind components, and zonal and meridional pseudostress components are evaluated and discussed.

The definition of cold event years

Figure 5-1 (c) presents an overlay of several time-series indices that can be used to define the existence of cold events in the tropical Pacific ocean. All of the time series have been filtered with a five-month triangle filter and normalized by their standard deviation. Two of these indices are based only on sea surface temperature anomalies (SSTA): Niño3.4 (the area average of SSTA from 170W to 120W, 4S to 4N) and EEQ (the area average of SSTA from 108W to 98W, 4S to 4N). Niño3.4 is a measure of the SST anomaly in the central equatorial Pacific, while EEQ is a measure of it in the eastern equatorial Pacific. A third index is the Troup Southern Oscillation Index (SOI), which is a normalized measure of the sea level pressure difference between Tahiti and Darwin, Australia. When time-filtered, it tends to be strongly negative during the years that are conventionally considered to be warm event years. To make this index match the other indices, the Troup SOI is presented here inverted (multiplied by -1). The fourth index (BEI, for Bjerknes ENSO Index) has recently been proposed specifically to identify the warm event years of ENSO (Harrison and Larkin, 1998a); no consideration of the characteristics of cold periods went into its construction. To the extent that the cold and warm cycles of ENSO are mirror images, we might expect the SOI and BEI to identify the same years as Niño3.4 or EEQ. Several things are revealed by a detailed look at Figure 5-1 (c).

First, the different measures of the condition of the tropical Pacific are correlated, during warm and cold events. But the correlation is not exact; cold conditions in the central equatorial Pacific do not always correspond to cold conditions in the eastern equatorial Pacific and positive swings of the SOI or negative swings of the BEI do not have a unique relationship to either of the SST anomaly indices. However, six cold years stand out (1949, 1955, 1970, 1973, 1975, 1988) from the others in that at least three of these indices agree that, during the period between May and the following April, there were three consecutive months with values greater than 1.25 σ. If the criterion is relaxed to be that two out of the four indices meet the above standard, then nine years (the above six plus 1950, 1954, and 1964) are identified. Applying the latter criterion, but of the opposite sign, identifies as warm event years: 1951, 1953, 1957, 1965, 1969, 1972, 1976, 1982, 1987, and 1991. Note that the separations between cold event years are much more uneven than between warm event years, and that cold event years can follow one another, which never happened with warm events in this period. Note also that the largest amplitude warm event years have maximum index values greater than 3–4 σ, while the largest cold event years have maximum index values of somewhat more than 2 σ.

The SSTA patterns of cold and warm events

Figure 5-2 (*c*) shows the composite (average) SST anomaly patterns associated with the nine cold events and the 10 warm events between 1946 and 1993 identified above. Only anomalies that are statistically significant at least at the 95 percent level are shown. Anomalies within the heavy dashed contours are significant at greater than 99 percent. We show every other month, beginning in September of the year before the event threshold was satisfied, which we denote Sep(−1), through July of the year after the threshold was satisfied (Jul(+1)). Cold event SST anomalies have been multiplied by (−1) to facilitate comparison of the warm and cold composite patterns.

Note that the cold and warm event patterns are very similar near the equator in the tropical Pacific, including the first appearance of significantly anomalous conditions off the west coast of South America, and the merging of a near-dateline anomaly with the eastern Pacific anomaly around Jul(0). The coldest water occurs more in the central equatorial Pacific than does the warmest water in warm events, and more of the tropical Pacific is significantly cold in cold events than is warm in warm events. The composite cold anomalies also persist longer into Yr(+1) than do the composite warm anomalies. Overall, it is fair to say that cold events are the roughly mirror image of warm events in their equatorial Pacific SST anomaly patterns. However, the patterns of surface wind anomalies and sea level pressure anomalies (not shown) are not so strongly inverses.

Outside the equatorial Pacific, substantial differences are clear. In particular, the SST anomalies along the west coast of North America, in the Indian Ocean, and the tropical North Atlantic are stronger during cold events. We note in passing that the composite anomalies based on the six-event subset of cold event years is similar in most of the large-scale patterns to the nine-event results presented here.

Composite results can be very useful, if the anomalies are typical of most events. We have examined the "typicality" of the composite features using time series taken from the area of the composite anomaly. Roughly two-thirds of the large-scale patterns shown in Figure 5-3 (*c*) satisfy a "typicality" criterion of being present, with a significant spatial extent, during more than half of the cold event years. There is no room here to summarize these results; they are fully presented in Larkin and Harrison (2001a).

Seasonal US weather anomalies associated with cold events

Figure 5-3 (*c*) presents a variety of results pertaining to the seasonal temperature and precipitation anomalies that have occurred during the nine cold event years since 1946. Results are presented for three seasons: Sept–Nov (Autumn), Dec–Feb (Winter) and Mar–May (Spring), moving left to right across the figure. Temperature anomaly results are presented in the upper three figure elements, with precipitation anomaly results just below. Three sets of six panels make up the figure.

The upper six panels show the nine cold event seasonal average weather anomalies. The second six panels show the first results after masking for statistical significance at the 80 percent level. Together these show that the largest anomalies are in winter and spring, with colder than normal (by $-1.6\,°C$ over a broad swath in spring, and extremes reaching $-2.4\,°C$) conditions in the upper US and warmer than normal (by more than $+1.6\,°C$) from the southeast up to the southern Great Lakes and westward down into Texas in winter. Scattered warmer than normal conditions in the southwest and Texas are found in autumn and spring. Precipitation anomalies tend to be small. They are largest along the Washington Cascades (mountains) in autumn and winter and in the Ohio River Basin in winter, where they can exceed 1"/mo. Smaller precipitation anomalies are found scattered across the southern parts of the US. No areas show drier than normal values larger than -1 inch/mo.

The final six panels of Figure 5-3 (*c*) show the number of times that weather anomalies in the most extreme quintiles occurred during cold event years. Having this occur four or more times is statistically significant at greater than the 90 percent level. The upper three panels indicate that the most extreme seasonal temperature anomalies generally have not occurred more than is likely to result from chance in autumn and winter, but that extreme cold anomalies in the upper US in spring are statistically significant. Precipitation anomalies are hard to summarize. The large wet anomalies in the northwest in autumn and winter are significant only on the western side of the Washington Cascades and in Idaho; the very wet conditions in the Ohio River basin are not significant. Dry conditions tend to be somewhat more likely to be significant than wet ones, but the results are not spatially consistent across many climate regions. The climate region results must be inspected in detail for any region of local interest.

Figure 5-4 (*c*) presents the same types of seasonal weather anomaly information as Figure 5-3, only for warm event seasons. In general, the weather associations are of the opposite sign to those associated with cold event seasons; warm in the northern US in winter and spring instead of cold; drier in the northwest instead of wet. However, the areas of statistically significant anomalies are rather different in the southern half of the US. Extreme anomalies are more strongly associated with warm event years than they are with cold event years.

Note

* This work was supported by NOAA's ERL (J. Rasmussen, Director), PMEL (E. Bernard, Director) and Office of Global Programs (M. Hall, Director). We wish to thank the entire PMEL/TMAP group. All of the analysis and figures of for this work were done using the Ferret data visualization program developed by NOAA/PMEL/TMAP (available at http://www.pmel.noaa.gov/ferret).

REFERENCES

Harrison, D.E. and N.K. Larkin, 1996: The COADS sea level pressure signal: A near global El Niño composite and time series view, 1946–93. *Journal of Climate*, 9(12): 3025–55.

Harrison, D.E. and N.K. Larkin, 1998a: El Niño-Southern Oscillation: Sea surface temperature and wind anomalies, 1946–1993. *Review of Geophysics*, 36(3): 353–400.

Harrison, D.E. and N.K. Larkin, 1998b: Seasonal U.U. temperature and precipitation anomalies associated with El Niño: Historical results and comparison with 1997–98. *Geophysical Research Letters*, 25(21): 3959–62.

Larkin, N.K. and D.E. Harrison, 2001: Tropical Pacific ENSO cold events, 1946–1995: SST, SLP and surface wind composite anomalies. *Journal of Climate* 14(19): 390–31.

Larkin, N.K. and D.E. Harrison, 2002: ENSO warm and cold event life cycles: Ocean surface anomaly patterns, their symmetries, asymmetries, and implications. *Journal of Climate* (in press).

Further thoughts on ENSO

Arun Kumar

Why categorize?

It seems that categorizing tropical Pacific sea surface temperature (SST) anomalies into discrete classes such as El Niño or La Niña events is necessary, because of the limited amount of historical data, the use of which allows the atmospheric impact for anomalous SSTs to be estimated. In principle, for continuous variations in the tropical Pacific SST anomalies, the atmospheric impacts should also form a continuous spectrum and, given a sufficient amount of observational data, can be estimated for all SST states (see Barnston, this volume). As an illustrative example, let us assume that the tropical Pacific SST anomalies have a single preferred spatial structure and that the differences in the ENSO events are only in terms of the amplitude of this spatial pattern alone. Let us further assume that the atmospheric response to these SST anomalies is also spatially fixed (e.g., the PNA [Pacific North American] pattern in the extratropics), and further, its amplitude is proportional to the amplitude of the anomalous SSTs (a similar conceptual model of atmospheric seasonal variability is discussed in Kumar and Hoerling, 2000).

Within this paradigm of continuous spectrum for the interannual SST variability, atmospheric impacts exist for all anomalous SST states. Starting from this premise one can still categorize the SST variability into three (or more) discrete classes, i.e., El Niño, La Niña, and Normal SST years and document corresponding atmospheric impacts. Inherent in this categorization is the assumption that for the Normal SST years the seasonal mean atmospheric climate state is also expected to be normal. On the other hand, if one follows the paradigm of continuous spectrum of interannual variability, one can argue that although the atmospheric impacts for the weak SST anomalies will be small, on average they will still be non-zero. For these small atmospheric impacts and forecasts based upon them, the societal impacts during individual events may also be of minimal benefit. However, depending on the users' tolerance to random outcomes, these small benefits can still accrue to be substantial over a large number of weak SST events. The concept of "benefit," therefore, is relative and depends on how long one is willing to use the seasonal forecast information to potentially benefit their cause. Following this argument, the concept of categorizing SST anomalies into discrete classes, to a certain extent, may undermine the potential usefulness of seasonal predictions. This is so, since all SST anomalies, weak or strong, have atmospheric impacts and decisions based on them, on average, should bring positive gains.

So why categorize? It is not just a matter of simplifying (or dumbing down) the language in which the seasonal forecasts are presented. The necessity to categorize is more fundamental and comes from the fact that we just do not have enough historical data to be able to infer the atmospheric signals for all SST states. Categorization allows us to pool "similar" anomalous SST states into discrete bins, and allows us to infer the atmospheric signals for these categories using a compositing technique. These composites can then be used for seasonal predictions, once it is known in which category the predicted future SSTs will fall.

The analogy of "loaded dice" and seasonal predictions

The analogy of a seasonal forecast as "loaded dice" is often given, but we should not forget the fact that one can toss the dice 100 times within the span of an hour, but can only "play" maybe 50 games based on seasonal prediction information in an individual's life span. Needless to say, the average length of the game over which one eventually benefits economically depends on how loaded the dice are. In terms of seasonal forecasts this translates into how strong the SST anomalies are; the larger SST anomalies, due to their larger impact on the seasonal mean state, make the dice "more loaded."

The potential utility of a probabilistic seasonal forecast is a hard thing to convey to individual users (e.g., individual farmers or fishermen or to the public at large). Human memory (or the human life span) is not long enough to clearly assess the average benefits from the use of seasonal predictions which can only accrue from the consistent application of seasonal forecast information over such a long period of time. One possible factor working against the potential use of seasonal forecast information is the inherent non-linearity or randomness in the economic benefits which can result from the decisions based on the seasonal forecasts. For example, use of seasonal prediction information and decisions based upon them may result in bankruptcy (if the prediction does not turn out to be correct), but may not result in riches (if the prediction is correct). The risk of such nonlinearity in outcomes and randomness of seasonal forecasts may prompt individual users to ignore altogether the seasonal predictions in their decision-making process (Nicholls, 1999).

Perhaps the biggest beneficiaries of seasonal predictions will be users such as large corporations. For them, the tolerance for random fluctuations from the use of seasonal prediction may be bigger and such large economic entities may be in a better position to play a larger number of games and be able to accrue the benefits from the use of seasonal predictions.

Do ENSO events increase weather-related disasters?

For both El Niño and La Niña events it will be interesting to study whether the impact of ENSO extremes is to increase the frequency of weather-related disasters, or whether the tropical SST events merely increase our ability to predict the geographical location where these disasters are more likely to occur. In other words, do ENSO events act like a lens which can focus "a random distribution of disasters" to particular geographical locations?

Why do we have so many SST forecasts?

During the La Niña Summit (Glantz, 1998), the proliferation of the SST forecasts was the cause of some concern. One can ask a question – why is it so? It is interesting to contrast this situation for the SST forecasts with the situation in the field of short-to-medium range atmospheric forecasts where prediction efforts are dominated solely by the large weather services. One fundamental difference between them is the use of initial condition information that, for the short to medium-range atmospheric forecasts, plays a crucial role. Gathering and analyzing information about the initial state of the atmosphere in a timely manner requires a large infrastructure, and hence, the lack of a multitude of weather forecasts. The importance of the initial conditions is dictated by the intrinsic nonlinearity of the atmospheric evolution. It may be that the proliferation of SST forecasts indicates that the oceanic predictions differ in this manner. If the evolution of the oceanic state is nonlinear and depends critically on its initial state, the SST predictions will eventually also become dominated by a few large organizations committed to assimilating and maintaining available ocean data analysis in a timely fashion. On the other hand, if the ocean dynamics is governed by the quasi-linear, low-frequency evolution starting from the knowledge of the large-scale structure of the initial oceanic anomalies, linear SST prediction methods may suffice (and proliferate) and will continue to be competitive with the predictions from the coupled (atmosphere-ocean) general circulation models.

Attribution of weather and climate events

During the La Niña Summit (Glantz, 1998) we argued endlessly about the attribution of extreme weather events to some underlying cause. Given that seasonal forecasts are probabilistic (even more so for the occurrence of extreme events on the sub-seasonal time scale), a basic question is whether it is of any practical benefit to seek attribution for individual extreme events. One can also ask whether, scientifically, it is even a well-posed problem? It is more meaningful to seek attribution in a statistical sense. For example, during El Niño winters, certain geographical regions over the west coast of the United States have a higher than normal probability for increased precipitation. This is likely to enhance the probability of either heavy rainfall events or increase their probability of frequency of occurrence. The attribution for individual rainfall events which can occur during any year may be a fruitless endeavor. From a probabilistic standpoint, attribution for individual extreme events also does not seem logical. A bottom line to judge practical advantages of attribution for individual events should be to ask the following question – will it help us to predict similar occurrences in the future?

REFERENCES

Glantz, M.H., 1998: *A Review of the Causes and Consequences of Cold Events: A La Niña Summit.* Executive summary of the workshop held 15–17 July 1998. Boulder, CO: National Center for Atmospheric Research.

Kumar, A. and M.P. Hoerling, 2000: Analysis of a conceptual model of seasonal climate variability and implications for seasonal prediction. *Bulletin of the American Meteorological Society*, 81, 255–64.

Nicholls, N., 1999: Cognitive illusions, heuristics, and climate prediction. *Bulletin of the American Meteorological Society*, 80, 1385–97.

Bridging the supply and demand gap in climate forecast production and use

Kenneth Broad

The La Niña Summit was intended primarily to assess the state of the art of physical science knowledge as well as the societal impacts of the cold phase of the ENSO cycle. However, several issues arose during the conference, which are directly relevant to the consideration of the societal applications of climate forecast information. This chapter touches on some of these topics and concludes with some suggestions for resolving the challenge of effective forecast dissemination.

The dissemination and utilization of seasonal-to-interannual climate forecasts involves two interrelated categories of action: the "supply side" and "demand side." "Supply side" refers to those agents and activities involved with producing and distributing climate information. The "demand side" refers to those agents (often referred to as end-users) who can potentially utilize climate information to make better informed decisions. Without a better understanding of the relationship between these aspects of this forecast equation, those concerned about the use of climate information will be handicapped in their efforts to effectively produce and disseminate relevant information, and in their ability to understand the socio-economic consequences created by providing such information.

The use of seasonal-to-interannual climate forecasts has for the most part been a result of supply-side driven efforts, with several national and international agencies keen to demonstrate what they perceived to be the societal value of their scientific achievements (e.g., National Oceanic and Atmospheric Administration, World Meteorological Organization). Mixed results of these efforts have led to the recognition that in order to produce and "market" useful and usable forecast products, more specific information would be needed about the end users, their informational needs, and the constraints on decision making that they face.

With few exceptions (e.g., McQuigg and Thompson, 1966; Glantz, 1977; Stewart et al., 1984), only recently has applied social science research focused on the demand side, and a range of socio-economic and political constraints to the optimal and equitable use of information for societal benefit have been identified (see e.g., Broad and Agrawala, 2000; Katz and Murphy, 1997; Mjelde et al., 1997; Orlove and Tosteson, 1999; Pfaff et al., 1999; Roncoli, 2000; Stern and Easterling, 1999). These studies have prompted a re-evaluation of the current policies and practices on climate forecast dissemination, and to the tailoring of information to the demands (i.e., needs) of end-users.

The focus of this chapter is on a set of related topics that bridge the supply and demand sides of the climate forecast problem – semantics, perception, and probability. These categories were chosen in an attempt to organize issues that emerged during discussions among the physical scientists at the La Niña Summit. This analysis is an attempt to provide some insight into the "culture" of the supply-side community (e.g., climate scientists), and to help better identify the obstacles to the improved use of climate information by a wide range of societal decision makers.

Semantics

It became clear from discussions during the Summit that many definitions and categorizations of ENSO events co-exist (for an example of this debate over the use of the term "El Niño," see Glantz and Thompson, 1981; Aceituno, 1992; Trenberth, 1997; Philander, 1998). This ambiguity is exaggerated in the case of La Niña, as there is no agreed-upon index for La Niña. In part this is because the climate anomalies influenced by La Niña are perceived in general to be of a lesser intensity. Within the scientific community, this ambiguity may be a healthy reality that encourages discussion, analysis, and debate. In fact, among this group of physical scientists, it may seem perfectly reasonable, as one of the Summit's speakers noted, that there appeared to be a "lingering El Niño and an incipient La Niña" occurring at the same time (July 1998). This ambiguity becomes problematic as it gets played out in public forums, however, where the media and decision makers are less versed in the scientific nuances that surround this seemingly simple scientific nomenclature.

Furthermore, official forecasts often include highly subjective terms such as "weak," "moderate," and "strong" which encompass the forecaster's knowledge of past events. This knowledge, however, is rarely made explicit to those on the receiving end of such information in a manner that allows an objective understanding of these terms (for examples and discussion of this issue, in the case of the 1997–98 El Niño, see Barnston et al., 1999). These definition-related ambiguities underlie some of the differing interpretations by producers of scientific information and by the users of this same set of information.

Scientific versus societal perceptions

The gap between perceptions of scientists and those of society extend beyond semantic differences. For example, scientific experts judge events by a relatively bounded set of parameters that can be measured objectively. Society in general, however, is less interested in a characterization of an event in terms of a change in a specific parameter, such as the depth of the thermocline along the equator. They will judge events by the severity of climate-related impacts in their particular geographic areas (e.g., flood, drought, disappearance of fish stocks). Thus, confusion arises as scientists categorize and predict large-scale environmental parameters in one part of the globe (e.g., sea surface temperatures in the central Pacific Ocean), and decision makers interpret these as predictions of impacts for application to their regions.

Because forecasts are made on a scale of hundreds of kilometers, even a "correct" forecast, which gives a good approximation of total rainfall or temperature averaged over a three-month period, will have temporal and spatial variations that will impact users differently. Agriculture, for instance, is sensitive not just to total rainfall, but to the timing of the onset of rains, mid-season droughts, and to extreme events such as severe storms and frosts, which are currently beyond our ability to predict several seasons in advance. This confusion between a "meteorological" forecast versus an "impacts" forecast is perhaps more acute in the case of La Niña, simply because we have collected less information on the impacts of La Niña events.

In attempting to explain impacts in general terms, misleading generalizations are often made. Summit participants discussed how La Niña's impacts are explained to the public by the media. They were portrayed simply as the opposite of El Niño, i.e., La Niña was expected to bring the opposite impacts to the same regions, even while scientific evidence does not support this generalization (Hoerling et al., 1997). Even the proposed impacts attributed to La Niña, such as the costly 1988 US Midwest drought, is still subject to debate. This uncertain explanation underscores the lack of understanding of the La Niña phenomenon.

Probability issues

As described above, aside from a few regions in the tropical Pacific, the forecasts of impacts that might have resulted from ENSO (from the point of view of many decision makers), have very limited skill due to their temporal and spatial imprecision. Because of the widespread and often uninformed attributions of many extreme weather events to El Niño during 1997–98, there is a risk of giving a false or misleading impression about the state of scientific knowledge of La Niña's impacts. As one speaker at the La Niña Summit put it: "once we imply attribution, we imply predictability." The scientific community is aware of the statistical insignificance of statements about impacts based on a sample of impacts during only two or three ENSO extreme events, or based on proxy reconstructions of ENSO impacts during past events. However, because the skill of how forecasts have performed in the past is not communicated clearly (if at all) to the public, decision makers have difficulty gauging the reliability of forecasts they are constantly being given.

Again, the media do not sell newspapers based on emphasizing uncertainty in their stories. They have displayed a constant tendency to sensationalize climate impacts, especially those related to El Niño (Glantz, 2001). In turn, policy makers are often pressured to respond, because of media hype related to El Niño and La Niña and its impacts. A lack of understanding of the uncertainty involved in both the forecasts of ENSO (specifically centered on the physical aspects of the central Pacific Ocean) and ENSO's teleconnections, can lead to poor policy making. The inherently probabilistic nature of climate forecasts thus becomes a central challenge to communicating information from the forecast community to the public, including policy makers.

Whose responsibility is it?

With regard to the supply side, scientists should not be expected to be educators of the public, media experts, or marketing specialists. Nevertheless, they can take actions to minimize the potential for misuse of climate forecasts and climate-related information by the public (this, of course, would be in addition to improving reliability of information and forecasts). From the perspective of the public, it is often difficult to distinguish among forecasts, predictions, and observations (e.g., Carr and Broad, 2000). A lack of transparency (e.g., honesty), about the level of skill associated with the climate forecasts has serious implications for optimal decision making and for developing trust by users in climate information. Someone surfing the Internet is as likely to find an experimental prediction from an unvalidated model, as they are to find a forecast which takes into account the output of several computer and statistical models, as well as information about other regional and local climatic factors in addition to ENSO.

Modelers are under pressure to highlight their successes and downplay their weaknesses. This raises a set of related questions:

- Should there be a mechanism for quality control of publicly distributed information?
- Should there be a clearinghouse for the numerous, often conflicting, climate forecasts?
- Should experimental results intended for sharing among scientific colleagues be password protected? Should forecast products approved by the forecast community be distributed to the public with the appropriate caveats?

An approach that addresses these concerns could minimize the burden on the media to correctly interpret scientific forecasts, e.g., relaying probability statements and scientific caveats to reduce the amount of conflicting information they produce (e.g., reducing the noise). This would minimize the possibility of misinterpretation and distortion of the intended forecast information. The International Research Institute for climate prediction (IRI) was created in 1996 to play such a role (iri.ldeo.columbia.edu). This organization straddles the gap between policy and science, drawing its incentives from and producing outputs for principals in both of these areas. Its relative success will be a reflection of how well the climate science community backs the goal to achieve a forecast production and delivery process (see Agrawala et al., 2001).

With regard to the demand side (i.e., the user community), only in the last few years has there been a notable increase in social science research geared towards an improved understanding of the impacts, potential uses, and sociopolitical constraints on the uses of climate information. Much of this increased activity has been accomplished in the context of El Niño research. In past years, it seems that only anecdotal evidence of La Niña's socio-economic effects was collected. Efforts should be intensified in this direction. Multidisciplinary research teams should include those who produce forecasts in order to identify the locations of robust (reliable) teleconnections and their socio-economic impacts on ecosystems and societies. The next logical step would be to evaluate current forecast capabilities within these regions in relation to the types of information that various decision makers need. Finally, in considering the application of forecasts to societal needs, consideration must be given to the possible negative and unintended consequences of introducing such information into the decision-making schemes, given the limited skill of the forecasts and the societal constraints on their use.

With such information, intermediary organizations bridging the gap between the supply and the demand side – whose mandate is to provide forecasts for the benefit of society (e.g., IRI, WMO, NOAA) – could use this information for the following purposes:
- To better inform decisions related to international and regional policies on climate forecast dissemination
- To identify the training and educational needs of potential user communities
- To identify ethical issues related to equitable forecast distribution (e.g., winners and losers)
- To identify the significant variables needed by decision makers to help guide basic research and the design of climate forecast products

As predictive skills, observational capabilities, and communication networks increase, there will be an improved synthesis of climate information. Hopefully, these forecast tools, informed by social science research, will lead to better prediction of the impacts that might occur throughout the ENSO cycle. They could also lead to more effective dissemination of useful climate information (including but not limited to forecasts) to affected societies. Such concerted efforts can eliminate some of the problems identified above, by encouraging societal capacity building. Doing so will help to shift the focus away from ad hoc responses to high-profile headlines that mention El Niño or La Niña. Instead, a foundation for societal resilience can be fostered through the development of effective coping mechanisms and strategies for adaptation over the long term. This would allow for better management of the impacts of seasonal-to-interannual climate variability.

REFERENCES

Aceituno, P. 1992: El Niño, the Southern Oscillation, and ENSO: Confusing names for a complex ocean atmosphere interaction. *Bulletin of the American Meteorological Society,* 73, 483–5.

Agrawala, S., K. Broad, and D.H. Guston, 2001: Integrating climate forecasts and societal decision making: Challenges to an emergent boundary organization. *Science, Technology, and Human Values* (in press).

Barnston, A.G., M.H. Glantz, and Y. He, 1999: Predictive skill of statistical and dynamical climate models in SST forecasts during the 1997–98 El Niño episode and the 1988 La Niña onset. *Bulletin of the American Meteorological Society,* 80, 217–43.

Broad, K. and S. Agrawala, 2000: The Ethiopia food crisis: Uses and limits of climate forecasts. *Science,* 289, 1693–4.

Carr, M.E. and K. Broad, 2000: Satellites, society, and the Peruvian fisheries during the 1997–98 El Niño. In D. Halpern (ed.), *Satellites, Oceanography and Society.* New York: Elsevier Science, 171–91.

Glantz, M.H., 1977: The social value of a reliable long-range weather forecast. *Ekistics*, 43, 305–13.

Glantz, M.H., 2001: *Currents of Change: El Niño and La Niña Impacts on Climate and Society*. Cambridge, UK: Cambridge University Press.

Glantz, M.H. and J.D. Thompson (eds.), 1981: *Resource Management and Environmental Uncertainty: Lessons from Coastal Upwelling Fisheries*. Vol. 11, *Advances in Environmental Science and Technology*. New York: John Wiley and Sons.

Glantz, M.H., D.G. Streets et al., 1998: *Exploring the Concept of Climate Surprises: A Review of the Literature on the Concept of Surprise and How It Is Related to Climate Change*. Argonne: US Department of Energy, Office of Energy Research.

Hoerling, M.P., A. Kumar, and M. Zhong, 1997: El Niño, La Niña, and the nonlinearity of their teleconnections. *Journal of Climate*, 9, 1769–86.

Katz, R.W. and A. Murphy (eds.), 1997: *Economic Value of Weather and Climate. Cambridge*. Cambridge: Cambridge University Press.

McQuigg, J.D. and R.G. Thompson, 1966: Economic value of improved methods of translating weather information into operational terms. *Monthly Weather Review*, 94, 83–7.

Mjelde, J.W., T.N. Thompson, C.J. Nixon, and P.J. Lamb, 1997: Utilizing a farm-level decision model to help prioritize future climate prediction research needs. *Meteorological Applications*, 4, 161–70.

Orlove, B.S. and J.L. Tosteson, 1999: *The Application of Seasonal to Interannual Climate Forecasts Based on El Niño-Southern Oscillation (ENSO) Events: Lessons from Australia, Brazil, Ethiopia, Peru, and Zimbabwe*. Berkeley, CA: Institute of International Studies, University of California, 1–60.

Pfaff, A., K. Broad, and M.H. Glantz, 1999: Who benefits from climate forecasts? *Nature*, 397, 645–6.

Philander, G.S., 1998: Who is El Niño? *EOS Transactions*, 79(13), 170.

Roncoli, M.C. (ed.), 2000: Anthropology and climate change: Challenges and contributions. Special issue of *Practicing Anthropology*, 22(4).

Stern, P.C. and W.E. Easterling (eds.), 1999: *Making Climate Forecasts Matter*. Washington, DC: National Academy Press.

Stewart, T.R., R.W. Katz, and A.H. Murphy, 1984: Value of weather information: A descriptive study of the fruit-frost problem. *Bulletin of the American Meteorological Association*, 65, 126–37.

Trenberth, K.E., 1997: The definition of El Niño. *Bulletin of the American Meteorological Society*, 78(12), 2771–7.

On the perception of probabilistic forecasts

Cecile Penland

Let's face it: black and white decisions are the easiest. That's why we like them. There's no judgment involved; we simply employ an "If ... Then ... Else ..." command as we would in a computer program and our decisions are made. This is why we demand a strict definition of El Niño, with an "Is it? Or isn't it?," when there clearly exists a continuum of magnitudes and patterns of sea surface temperatures (SSTs). A bounded set of Indo-Pacific SSTs, be it ever so continuous, must have its maximum and minimum values, and we call these El Niño and La Niña, respectively, but the terms also apply to a range of values neither so warm nor so cold as the extremes. Where, then, does one draw the line when a natural delineation does not present itself? (see Kumar, this volume).

The same quandary occurs in considering the patterns of SSTs. We know a strong El Niño pattern when we see it, but there are variations of this pattern which may also be considered El Niño, variations which continuously meld into other patterns until we are not quite sure where to say that the El Niño pattern started or stopped. How do we handle this? We may declare that a strict definition is needed, but how does one apply "on-off" logic to a continuous variable? It seems that an equivalent question is as follows: how do we force a physical phenomenon, over which we have little or no control, to fit our methods of contemplation? When faced with the question put this way, it seems easier to change the way we think, particularly if we flatter ourselves that we do have some small control over our own thought processes. But there is a price:

changing the way we think about a continuous, complex system requires us to admit the concepts of uncertainty and dynamical probability.

Many scientists claim their willingness to consider probabilistic description of the El Niño phenomenon. The demand for yes-no answers is blamed on the general public, who, it is claimed by some scientists and science writers, are incapable of understanding or accepting complex systems or probabilistic forecasts. Is this claim accurate? Well, most people shrug off the impossibility of describing with certainty what any person will do, but they get married anyway. A volatile stock market is still a popular place to invest pension funds. One of the most highly populated areas in the country (California) is plagued with earthquakes. People seem willing to make generalizations based on statistics and to accept the impossibility of predicting complex systems with infinite accuracy, as long as they are convinced that the system is truly complex, and as long as the decision-making process isn't perceived as "science."

I believe that public perception of scientists as chosen cognoscenti, and many scientists' willingness to accept that perception, has done more to hinder the communication between scientists and nonscientists than any other cause. A popular ad for a mortgage broker showed the stereotypical scientist – white male, portly, fifty-ish, turtleneck sweater, white hair and moustache, holding a pipe – with the caption "You don't have to be a rocket scientist to understand our mortgage programs." Message: "You can't understand what scientists do, but you can understand all of the risks and legal ramifications of what may be the largest, most important financial transaction of your life." No wonder the public is willing to believe itself incapable of understanding any but the simplest scientific concepts! (Side comment on ads: It is interesting how little of the discussion of the hype about El Niño by "the media" separates the news media from the advertising media.)

But, given the evidence that nonscientists make probabilistic choices every day, why do scientists doubt the public ability to accept probability? In my opinion, many physical scientists themselves are uncomfortable with the role probability plays in modern scientific research. This discomfort is understandable. After all, most of our knowledge of classical physics comes from a reductionist philosophy of science coupled with carefully controlled experiments. The key word here is "control." An uncontrolled experiment implies an incompetent scientist in the traditional scientific culture, and this cultural wisdom has historically been correct in laboratory experiments where specified uncertainty does not indicate a lack, but rather the extent, of control. But how, for example, can we deterministically describe the collective behavior of molecules in a macroscopic container of gas? We can't, and that is how dynamical probability theory found its way into the physical sciences only one century ago. But the mere necessity of considering dynamical probability indicates a lack of control, somehow, and while most scientists can intellectually accept the necessity for probabilistic dynamical description to some extent, I fear that the emotional difficulties that the concept of dynamical probability presents to the traditional scientist keeps him or her from wholeheartedly accepting the concept as a physical phenomenon. It is difficult to accept fully a concept that is emotionally repugnant, and I believe it is this difficulty which leads many physical scientists to consider dynamical probability "hard." If we ourselves have trouble dealing with it, we believe it must be hard for the general public to understand.

I submit that the general public has never been given an adequate opportunity to deal with probabilistic forecasts. Without even considering the effect of the absolutely certain-sounding predictions being made in most countries, would the unpopular probabilistic forecasts of the 1997–98 El Niño issued in Australia have had more acceptance if all forecasts, not just the El Niño forecasts, allowed for a sense of uncertainty? Imagine a television meteorologist (using the metric system): "Well, folks, highs in the low 30s plus or minus about 6 degrees and lows at night in the high teens plus or minus about 4 degrees, so it looks nice and warm. But remember, that's quite a bit of spread!" One cannot help but ask whether or not confused interpretations of probabilistic forecasts are not a consequence of a simple lack of practice interpreting probabilistic forecasts by the general public, a lack of communication skills by scientists, or a combination of both. After all, there seems to have been little or no backlash in Florida against forecasters who frankly admitted that they were not sure where or if Hurricane Georges was going to hit the southern coast but provided a range of possibilities (nonscientist friends in Florida, 1998, personal communication). I honestly don't think the communication problem lies with the public. I think the problem lies with the inability scientists have to come to terms with the fact that there are physical reasons why we must sometimes be unsure, and that it's okay to let other people know it.

Appendix: Frequently asked questions (FAQ) about La Niña and El Niño from NOAA's website

www.elnino.noaa.gov/lanina_new_faq.html

What is La Niña?

La Niña is defined as cooler than normal sea surface temperatures in the tropical Pacific ocean that impacts global weather patterns. La Niña conditions may persist for as long as two years.

What's the difference between La Niña and El Niño?

El Niño and La Niña are extreme phases of a naturally occurring climate cycle referred to as El Niño/Southern Oscillation. Both terms refer to large-scale changes in sea surface temperature across the eastern tropical Pacific. Usually, sea surface readings off South America's west coast range from the 60s to 70s F, while they exceed 80 degrees F in the "warm pool" located in the central and western Pacific. This warm pool expands to cover the tropics during El Niño, but during La Niña, the easterly trade winds strengthen and cold upwelling along the equator and the west coast of South America intensifies. Sea surface temperatures along the equator fall as much as 7 degrees F below normal.

Why do El Niño and La Niña occur?

El Niño and La Niña result from interaction between the surface of the ocean and the atmosphere in the tropical Pacific. Changes in the ocean impact the atmosphere and climate patterns around the globe. In turn, changes in the atmosphere impact the ocean temperatures and currents. The system oscillates between warm (El Niño) to neutral (or cold La Niña) conditions with an average roughly every 3–4 years.

What are the global impacts of La Niña?

Both El Niño and La Niña impact the global and US climate patterns. In many locations, especially in the tropics, La Niña (or cold episodes) produces the opposite kinds of climate variations from El Niño. For instance, parts of Australia and Indonesia are prone to drought during El Niño but are typically wetter than normal during La Niña.

What are the US impacts of La Niña?

La Niña often features drier than normal conditions in the southwest in late summer through the subsequent winter. Drier than normal conditions also typically occur in the central plains in the fall and in the southeast in the winter. In contrast, the Pacific Northwest is more likely to be wetter than normal in the late fall and early winter with the presence of a well-established La Niña. Additionally, on average La Niña winters are warmer than normal in the Southeast and colder than normal in the Northwest.

Does a La Niña typically follow an El Niño?

No, a La Niña episode may but does not always follow an El Niño.

Is there such a thing as "normal," aside from El Niño and La Niña?

Over the long-term record, sea surface temperatures in the central and eastern tropical Pacific diverge from normal in a roughly bell-curve fashion, with El Niño and La Niña at the tails of the curve. Some researchers argue there are only two states, El Niño and non-El Niño, while others believe either El Niño or La Niña is always present to a greater or lesser degree. According to one expert, NCAR's Kevin Trenberth, El Niño events were present 31 percent of the time and La Niña events 23 percent of the time from 1950 to 1997, leaving about 46 percent in a neutral state. The frequency of El Niño events has increased in recent decades, a shift being studied for its possible relationship to global climate change.

How often does La Niña occur?

El Niño and La Niña occur on average every 3–5 years. However, in the historical record the interval between events has varied from 2–7 years.

According to the National Centers for Environmental Prediction, this century's previous La Niñas began in 1903, 1906, 1909, 1916, 1924, 1928, 1938, 1950, 1954, 1964, 1970, 1973, 1975, 1988, and 1995. These events typically continued into the following spring. Since 1975, La Niñas have been only half as frequent as El Niños.

How long does a La Niña last?

La Niña conditions typically last approximately 9–12 months. Some episodes may persist for as long as two years.

How do scientists detect La Niña and El Niño and predict their evolution?

Scientists from NOAA and other agencies use a variety of tools and techniques to monitor and forecast changes in the Pacific Ocean and the impact of those on global weather patterns. In the tropical Pacific Ocean, El Niño is detected by many methods, including satellites, moored buoys, drifting buoys, sea level analysis, and expendable buoys. Many of these ocean observing systems were part of the Tropical Ocean Global Atmosphere (TOGA) program, and are now evolving into an operational El Niño/Southern Oscillation (ENSO) observing system. NOAA also operates a research ship, the Ka'Imimoana, which is dedicated to servicing the Tropical Ocean Atmosphere (TAO) bouy network component of the observing system.

Large computer models of the global ocean and atmosphere, such as those at the National Centers for Environmental Prediction use data from the ENSO observing system as input to predict El Niño. Other models are used for El Niño research, such as those at NOAA's Geophysical Fluid Dynamics Laboratory, at Center for Ocean-Land-Atmosphere Studies, and other research institutions.

Why is predicting these types of events so important?

Better predictions of the potential for extreme climate episodes like floods and droughts could save the United States billions of dollars in damage costs. Predicting the onset of a warm or cold phase is critical in helping water, energy and transportation managers, and farmers plan for, avoid, or mitigate potential losses. Advances in improved climate predictions will also result in significantly enhanced economic opportunities, particularly for the national agriculture, fishing, forestry, and energy sectors, as well as social benefits.

What is the relationship between El Niño/La Niña and global warming?

The jury is still out on this. Are we likely to see more El Niños because of global warming? Will they be more intense? These are the main research questions facing the science community today. Research will help us separate the natural climate variability from any trends due to man's activities. We cannot figure out the "fingerprint" of global warming if we cannot sort out what the natural variability does. We also need to look at the link between decadal changes in natural variability and global warming. At this time we can't preclude the possibility of links but it would be too early to definitely say "yes, there is a link."

Is this a "La Niña" hurricane/tropical storm/drought/fire/flood/ winter storm?

It is inaccurate to label individual storms or events as a La Niña or El Niño event. Rather, these climate extremes affect the position and intensity of the jet streams, which in turn affect the intensity and track of storms. During La Niña the normal climate patterns are enhanced. For examples in areas that would normally experience a wet winter, conditions would likely be wetter than normal.

How is La Niña influencing the Atlantic and Pacific hurricane seasons?

Dr. William Gray at the Colorado State University has pioneered research efforts leading to the discovery of La Niña impacts on Atlantic hurricane activity, and to the first and presently only operational long-range forecasts of Atlantic basin hurricane activity. According to this research, the chances for the continental US and the Caribbean Islands to experience hurricane activity increases substantially during La Niña.

What impacts do El Niño and La Niña have on tornadic activity across the country?

Since a strong jet stream is an important ingredient for severe weather, the position of the jet stream determines the regions more likely to experience tornadoes. Contrasting El Niño and La Niña winters, the jet stream over the United States is considerably different. During El Niño the jet stream is oriented from west to east over the northern Gulf of Mexico and northern Florida. Thus this region is most susceptible to severe weather. During La Niña the jet stream extends from the central Rockies east-northeastward to the eastern Great Lakes. Thus severe weather is likely to be further north and west during La Niña than El Niño.

How are sea surface temperatures monitored?

Sea surface temperatures in the tropical Pacific Ocean are monitored with data buoys and satellites. NOAA operates a network of 70 data buoys along the equatorial Pacific that provide important data about conditions at the ocean's surface. The data is complimented and calibrated with satellite data collected by NOAA's Polar Orbiting Environmental Satellites, NASA's TOPEX/POSEIDON satellite and others.

How are the data buoys used to monitor ocean temperatures?

Observations of conditions in the tropical Pacific are essential for the prediction of short term (a few months to one year) climate variations. To provide necessary data, NOAA operates a network of buoys that measure temperature, currents, and winds in the equatorial band. These buoys transmit data that are available to researchers and forecasters around the world in real time.

Why has the public not heard much about La Niña before now?

For many decades, scientists have known about the oscillation in atmospheric pressure across the tropical Pacific at the heart of both El Niño and La Niña. However, La Niña's effects on fisheries along the immediate coast of South America, where El Niño was named, are benign rather than destructive, so La Niña received relatively little attention there. Research on La Niña increased after its wider impacts (often called teleconnections) were recognized in the 1980s.

Where can I find more information on La Niña?

The Internet is the greatest source of information on El Niño, La Niña, and weather and climate data. NOAA has created one primary website that allows you to link to many other resources: www.elNiño.noaa.gov/laNiña.

- Specific information on La Niña predictions and other background is available NOAA's Climate Prediction Center at: nic.fb4.noaa.gov
- Information on NOAA's latest research initiatives is available at from the Climate Diagnostic Center at: www.cdc.noaa.gov/ENSO/
- NOAA's Pacific Marine Environmental Laboratory also has lots of valuable data including current observations from the network of data buoys in the tropical Pacific Ocean: www.pmel.noaa.gov/toga-tao/el-Nino/

Contributors

Anthony G. Barnston, International Research Institute, Lamont-Doherty Earth Observatory, Palisades, New York, USA

Guillermo J. Berri, Department of Atmospheric Sciences, University of Buenos Aires, Argentina

Kenneth Broad, University of Miami, Rosenstiel School of Marine and Atmospheric Science, Miami, Florida, USA

Allan D. Brunner, National Aeronautics and Space Administration, Goddard Space Flight Center, Greenbelt, Maryland, USA

Antonio J. Busalacchi, National Aeronautic and Space Administration, Goddard Space Flight Center, Greenbelt, Maryland. Now at Earth System Science Interdisciplinary Center, University of Maryland, College Park, Maryland, USA

M. Pilar Cornejo-Grunauer, Facultad de Ing. Maritima y Ciencias, Guayaquil, Ecuador

Eduardo A. Flamenco, Instituto Nacional de Tecnología Agropecuaria, Castelar, Argentina

Ray Garnett, Geography Department, University of Saskatchewan, Saskatoon, Saskatchewan, Canada (retired)

Michael H. Glantz, National Center for Atmospheric Research, Environmental and Societal Impacts Group, Boulder, Colorado, USA

Nicholas E. Graham, Hydrologic Research Center, 12780 High Bluff Drive, San Diego, California, USA

Charles "Chip" Guard, National Weather Service Forecast Office, Barrigada, Guam

D.E. Harrison, Joint Institute for the Study of the Atmosphere and Ocean, University of Washington, Seattle, Washington, USA

Kzi Hasan Imam, Bangladesh Public Administration Training Centre, Savar, Dhaka, Bangladesh

Martin Hoerling, National Oceanic and Atmospheric Administration, Cooperative Institute for Research in Environmental Sciences, Climate Diagnostic Center, Boulder, Colorado, USA

Monjurul Hoque, Bangladesh Public Administration Training Centre, Savar, Dhaka, Bangladesh

Ekram Hossain, Bangladesh Public Administration Training Centre, Savar, Dhaka, Bangladesh

Rafael Hurtado, Faculty of Agronomy, University of Buenos Aircs, Argentina

John L. Kermond, National Oceanic and Atmospheric Administration, Office of Global Programs, Silver Spring, Maryland, USA

Tahl Kestin, Cooperative Centre for Southern Hemisphere Meteorology, Monash University, Clayton, Victoria, Australia. Now at International Research Institute, Lamont-Doherty Earth Observatory, Palisades, New York, USA

George Kiladis, National Oceanic and Atmospheric Administration, Aeronomy Lab, Boulder, Colorado, USA

Kamal Kishore, Asian Disaster Preparedness Center, Klong Luang, Pathumthani, Thailand

R.H. Kripalani, Indian Institute of Tropical Meteorology, Pune, India

Ashwini Kulkarni, Indian Institute of Tropical Meteorology, Pune, India

Arun Kumar, National Oceanic and Atmospheric Administration, National Centers for Environmental Prediction, Climate Modeling Branch, Environmental Modeling Center, Camp Springs, Maryland, USA

Christopher W. Landsea, National Oceanic and Atmospheric Administration, Atlantic Oceanographic and Meteorological Laboratory, Hurricane Research Division, Miami, Florida, USA

Sim Larkin, Joint Institute for the study of the Atmosphere and Oceans (JISAO), University of Washington, Seattle, Washington, USA

Mark Majodina, Research Group for Seasonal Climate Studies, South African Weather Bureau, Pretoria, South Africa

Nathan Mantua, Joint Institute for the Study of the Atmosphere and Oceans, University of Washington, Seattle, Washington, USA

Michael J. McPhaden, National Oceanic and Atmospheric Administration, Pacific Marine Environmental Laboratory, Seattle, Washington, USA

Gerald Meehl, National Center for Atmospheric Research, Climate and Global Dynamics Division, Boulder, Colorado, USA

Mikiyasu Nakayama, United Graduate School of Agricultural Science, Tokyo University of Agriculture and Technology, Tokyo, Japan

Lino Naranjo-Diaz, Climate Center, Institute of Meteorology of Cuba, Havana, Cuba. Now at Universidade de Santiago de Compostela, Departamento de Física, Santiago, Spain

James O'Brien, Center for Ocean-Atmospheric Prediction Studies, Florida State University, Tallahassee, Florida, USA

Norma Ordinola, Laboratorio de Fisica, Universidad de Piura, Piura, Peru

Cecile Penland, National Oceanic and Atmospheric Administration, Cooperative Institute for Research in Environmental Sciences, Climate Diagnostics Center, Boulder, Colorado, USA

Roger A. Pielke, Jr., National Center for Atmospheric Research, Environmental and Societal Impacts Group. Now at the Center for Science and Technology Policy Research, University of Colorado, Cooperative Institute for Research in Environmental Sciences, Boulder, Colorado, USA

Maurice Roos, California Department of Water Resources, Sacramento, California, USA

Amir Shabbar, Meteorological Service of Canada, Environment Canada, Downsview, Ontario, Canada

Syed Shamsul Alam, Bangladesh Public Administration Training Centre, Savar, Dhaka, Bangladesh

Gary Sharp, Centre for Climate/Ocean Resources Study

Liliana Spescha, Faculty of Agronomy, University of Buenos Aires, Argentina

A.R. Subbiah, Asian Disaster Preparedness Center, Klong Luang, Pathumthani, Thailand

Raúl A. Tanco, Department of Geophysics, University of La Plata, Argentina

Kevin E. Trenberth, National Center for Atmospheric Research, Climate and Global Dynamics Division, Boulder, Colorado, USA

Joseph Tribbia, National Center for Atmospheric Research, Climate and Global Dynamics Division, Boulder, Colorado, USA

Peter E.O. Usher, United Nations Environment Programme, Nairobi, Kenya (retired)

Wang Shao-wu, Peking University, Beijing, China (retired)

Wei Gao, Cooperative Institute for Research in the Atmosphere, Colorado State University, Fort Collins, Colorado, USA

Tsegay Wolde-Georgis, Embassy of Ethiopia, Washington, DC, USA

Warren Wooster, University of Washington, School of Marine Affairs, Seattle, Washington, USA

Stephen Zebiak, International Research Institute, Lamont-Doherty Earth Observatory, Palisades, New York, USA

Index

Catalogue Request

Name: _____

Address: _____

_____

Tel: _____

Fax: _____

E-mail: _____

To receive a catalogue of UNU Press publications kindly photocopy this form and send or fax it back to us with your details. You can also e-mail us this information. Please put "Mailing List" in the subject line.

United Nations University Press

53-70, Jingumae 5-chome
Shibuya-ku, Tokyo 150-8925, Japan
Tel: +81-3-3499-2811   Fax: +81-3-3406-7345
E-mail: sales@hq.unu.edu   http://www.unu.edu